J.C.E. Underwood

Introduction to Biopsy Interpretation and Surgical Pathology

Second Edition

With 85 Figures

Springer-Verlag
London Berlin Heidelberg New York
Paris Tokyo

J.C.E. Underwood, MD, FRCPath.

Joseph Hunter Professor of Pathology, University of Sheffield Medical School, and Honorary Consultant Histopathologist, Royal Hallamshire Hospital, Sheffield S10 2RX, England

ISBN-13: 978-3-540-17495-0   e-ISBN-13: 978-1-4471-1473-4
DOI: 10.1007/978-1-4471-1473-4

British Library Cataloguing in Publication Data
Underwood, J.C.E.
Introduction to biopsy interpretation and surgical pathology.—2nd ed.   1. Histology, Pathological—Technique
I. Title 616.07'583   RB43

Library of Congress Cataloging-in-Publication Data
Underwood, J.C.E. (James Cressee Elphinstone)
Introduction to biopsy interpretation and surgical pathology.
Includes bibliographies and index.
1. Pathology, Surgical.   2. Biopsy.   I. Title
[DNLM: 1. Biopsy.   2. Diagnosis, Surgical. WO 142 U55i]
RD57.U52   1987   617'.0758   87–9510

Filmset and printed by BAS Printers Limited, Over Wallop, Hampshire.

2128/3916–543210

Preface

Preface to the Second Edition

Since publication of the first edition, continuing developments in histopathology have motivated the inclusion of four new chapters: cytology, immunohistology, quality control and assessment, and the autopsy. The increasing use of cytology in clinical practice and advances in techniques and diagnostic interpretation justify the devotion of more space than formerly to this important topic. Immunohistology now merits a separate chapter because it has a status beyond that of being just another "special stain" and there are certain aspects of technique and interpretation that are peculiar to it. Quality control and assessment in histopathology are very topical and important now that greater attention is being paid to cost effectiveness and the auditing of many aspects of clinical practice. The autopsy is an integral part of the professional life of a histopathologist and, though it lacks the immediate clinical impact of the biopsy diagnosis, it nevertheless constitutes an important activity in all hospitals; without detailing dissection techniques I have provided some information and guidance about the autopsy which I believe will be useful to trainees and of interest to others. The remainder of the book has been thoroughly revised and updated, and illustrations and tables added where experience has shown that the text is insufficient.

Sheffield, October 1986   J.C.E. Underwood

Preface to the First Edition

This book is primarily addressed to the needs of the trainee histopathologist. It is intended to bridge that gap between the descriptive histopathology taught as part of the undergraduate medical curriculum and the interpretative skills required of the diagnostician. My object is to convey the basic general principles, in theory and practice. Books are, however, only adjuncts to practical experience and not substitutes for it. Indeed, to obtain the maximum benefit from this book it is essential that the reader is actively involved in the work of a diagnostic laboratory. Only in this way can the trainee become thoroughly conversant with the rudiments of biopsy interpretation.

To retain detailed knowledge of the histological appearances of the plethora of diseases and their permutations to which humans are subject is beyond the mental resources of most individuals. For this reason, histopathologists probably refer to books more often than do most other specialists. This book aims to provide a core of knowledge sufficient to master the fundamental aspects, while still encouraging the intelligent use of all those indispensable atlases, monographs and fascicles for which there is no substitute.

I make no apologies for beginning the text with a brief account of the history of the biopsy. The origins of biopsy diagnosis might be dismissed as having little relevance to current practice, but it is often easier to understand a subject if one can see how it has evolved and recognise the various ways in which it may continue to develop. Also, it would have been easy to evade the need for a systematic consideration of the actual processes of interpretation. Having decided that the reader should be introduced to heuristics and pattern recognition, I was anxious to avoid giving an idiosyncratic account of something so personal and subjective. I hope I have succeeded in showing that there is nothing particularly arcane about the methods we use to interpret histological images.

The histopathological diagnosis does not begin and end with the inert, two-dimensional, brightly coloured artifact of the stained histological section. It begins with a living patient and ends with a credible report embodying a diagnosis or any other information that may guide the treatment of the patient. Modern investigative techniques may be needed to solve some diagnostic problems and for that reason I have included a general introduction to newer methods of histopathological analysis of proven clinical value. These include electron microscopy, immunological methods and stereology. Those who work in laboratories without such resources will, I hope, find guidance to the potential usefulness of these relatively sophisticated techniques. Perhaps they will be prompted to seek appropriate help and advice from colleagues who have access to the materials and skills that are necessary to solve some of the more difficult problems that may come their way.

I have referred to published work liberally but not uncritically. I hope that the text will whet the reader's appetite for more information and give directions as to where it may be satiated.

For the trainee, this will be an introduction to biopsy interpretation and surgical pathology in the context of contemporary clinical medicine. The text is an overview and commentary on current theory and practice in histopathology; I would, therefore, be well pleased if senior pathologists or clinicians also found something of interest and value in these pages.

Sheffield, January 1981   J.C.E. Underwood

Acknowledgements

I am very grateful to the many readers of the first edition who made helpful and positive comments, most of which I have been able to accommodate. Although the book continues to be aimed primarily at trainee histopathologists, it has been read by many of their more experienced seniors and I am indebted to them for their encouragement and advice in the preparation of this new edition.

I continue to be indebted to Drs. M. E. Boxer, C. D. Franklin, J. R. Goepel and L. Harvey, whose guidance and comments during the preparation of the first edition was invaluable. The assistance of Mr. T. E. Durrant and Mr. T. Gray with aspects of electron microscopy and X-ray microanalysis is acknowledged, and Mr. G. Anderson's advice about special stains has continued to be indispensable. Dr. N. Rooney and Miss Patricia Mills helped with the selection of illustrative material for the new chapter on immuno-histology, and Mr. J. Lawry provided the DNA profiles illustrating the section on flow cytometry.

Many colleagues have allowed me to use illustrations prepared from their own case material, notably Drs. A. J. Coup, J. R. Goepel, P. Gray, A. S. Hill, M. A. Parsons, C. M. D. Ross and J. R. Shortland.

Permission to reproduce illustrations from their published work or unpublished material was given generously by Dr. J. D. Davies of Bristol, Dr. B. Dixon of Leeds, Professor I. R. H. Kramer of London and Dr. E. J. Wilkinson of Milwaukee. I am also grateful for permission from the Editor and publishers of the Lancet to reproduce extended quotations from the published works of J. H. Bennett and Ernest Shaw; the Editor of the American Journal of the Medical Sciences and Charles B. Slack, Inc. to reproduce the extended quotation from the published work of F. Donaldson; the Editor of the British Journal of Cancer and H. K. Lewis & Co. Ltd. to reproduce Fig. 1.1; Dr. J. E. Rhoads, Editor, Cancer, and the American Cancer Society to reproduce Figs. 2.4 and 3.13; Professor R. E. Cotton, Editor, Histopathology, and Blackwell Scientific Publications Ltd. to reproduce Fig. 9.8; and the Editors and publishers of the Journal of Clinical Pathology to reproduce Fig. 10.6. Figure 9.1 was kindly supplied by Philips of Eindhoven and Figure 10.3 is modified from an illustration provided by Becton Dickinson.

The quality of the illustrations is to the credit of Mr. M. J. Eaton. My secretary, Mrs. Mary Hogg, and latterly Mrs. Brenda Barrass, deciphered and typed and sometimes retyped the script; for this and all their other help with the preparation of this edition, I am most grateful. The enthusiasm and guidance of Mr. Michael Jackson and his colleagues of Springer-Verlag continues to be much appreciated.

This new edition would not have appeared without the personal support of my closest friends and family. I dedicate it to them.

Contents

Explanatory Note

Some clarification of the words "biopsy", "surgical pathology" and "histopathology" is necessary at this point.

"Biopsy" is derived from the Greek βίοδ (life) and ὄψιδ (view). The word is used in two ways; first, to describe the process of removal of tissue from the living patient for diagnostic purposes (e.g. to biopsy the liver) and, secondly, to describe the actual tissue that has been removed (e.g. a liver biopsy). In both senses the emphasis is on diagnosis. "Biopsy" is also loosely used to describe the removal of any piece of tissue from the living body for examination; the examination may just be an assessment of a disease process that has been previously diagnosed (e.g. gastrectomy for endoscopically diagnosed carcinoma).

"Surgical pathology" is a branch of anatomical pathology dealing with the examination of tissues for the diagnosis and guidance in the care of patients; this safely encompasses the strict and loose meanings of "biopsy". At the risk of being pedantic, it must be pointed out that "surgical pathology" is restricted by some dictionaries to the pathology of those diseases accessible to surgical procedures.

"Histopathology" is the study of diseased tissue; strictly used, this applies only to the cytological and histological structure of the tissues, though some might prefer a broader definition that includes any investigation of diseased tissues. "Diagnostic histopathology" is the application of histopathology to the examination of tissues removed for the diagnosis and care of patients.

1　Diagnostic Histopathology

In the twentieth century histopathology has developed into a major branch of clinical medicine. While surgeons and physicians continue to bear direct responsibility for the provision of clinical services, investigative and diagnostic specialists are now an integral part of the team that cares for patients. Histopathologists are increasingly involved in work that has a fundamental bearing on the immediate management of patients and are frequently party to therapeutic decisions. The modern histopathologist therefore requires both an intimate knowledge of biopsy appearances and the ability to interpret biopsies in the context of contemporary clinical practice. Before examining the various ways in which the histopathological interpretation of biopsies and surgical resections contributes to clinical practice, the origins of the discipline merit brief recapitulation.

Haruspicy, originating in ancient Babylon, was the art of foretelling the future by studying the organs of slaughtered animals. Some cryptic message latent within the entrails was translated by the haruspex and broadcast to the people (King and Meehan 1973). The haruspex and the diagnostic histopathologist both prognosticate on information obtained from the examination of tissues. They differ only as do fancy and fact.

Origins of Histopathology

The history of histopathology can be divided into two phases. First, the simple recognition and description of the morbid changes in tissues associated with disease. Second, the use of such knowledge to identify a disease and to predict its behaviour and susceptibility to treatment in a living subject.

Claudius Galen (130–200) is generally acknowledged to be the first to have given detailed descriptions of the structural changes in the body associated with disease. He constructed a classification of tumours, lesions he attributed to an excess of black bile. In Italy, Marco Aurelio Severino (1580–1656) and Giovanni Battista Morgagni

(1682–1771) pioneered the renaissance of morbid anatomy. Severino, in his illustrated book *De Recondite Abscessorum Natura* (1632), classified breast cancer into four different types. However, Morgagni is usually regarded as the founder of modern pathological anatomy. His work, based on a meticulous correlation of the clinical history and autopsy findings, was not merely descriptive but a genuine attempt to understand disease processes. The status of pathology as an independent science was established by Matthew Baillie (1761–1823) with the publication in London of *The Morbid Anatomy of Some of the Most Important Parts of the Human Body* (1793). An atlas followed a few years later. But it was not until the microscope was applied to the study of diseased tissues that information of potentially diagnostic value was obtained.

Thin sections, necessary for microscopy, could be cut only if the tissue was hardened in some way to prevent deformation. Freezing was simple and effective; wax embedding was still many years away. Sir Everard Home (1763–1832) published pictures of the first histological sections of tumours in this book *A Short Tract on the Formation of Tumours* (1830), but derived few conclusions from them. In Germany, Johannes Müller (1801–1858) in *Über den Feinern Bau und die Formen der Krankhaften Geschwülste* (1838) was able to distinguish different tumours by microscopy.

Among the earliest descriptions of the use of microscopy in the actual diagnosis of tumours and ulcers are those of Bennett (1845) working in Edinburgh, Scotland, and Donaldson (1853) of Baltimore, Maryland. They were the first to show that therapeutically useful information could be obtained from the microscopic examination of tumours and tumour-like lesions. Though they used smears rather than sections, Bennett and Donaldson were largely responsible for transforming human pathology from a purely descriptive discipline into an entirely novel diagnostic method. Their enthusiastic efforts mark the birth of diagnostic histopathology and cytology.

Another woman, 50 years of age, of cachectic appearance, had for six months an ulcer in the left breast. It was about an inch from the nipple, sunk deep into the substance of the organ, and was about the size of a walnut. Its edges and the surrounding substances were firm and indurated. The glands of the axilla were slightly enlarged. The right breast was healthy. It became a point to determine whether the ulcer was malignant or simple; whether an operation was or was not to be resorted to? An examination of the fluid upon the surface of the ulcer, with the microscope, exhibited—1st. Pus cells, which, on the addition of acetic acid, presented the usual granular nucleus; 2nd. There were several flat scales, presenting all the character of pavement epithelium. 3rd. Were cells of an elongated form, similar to those observed in granulations, and cellular tissue in an early stage.

From these circumstances it was diagnosed that the ulcer was not malignant, and it subsequently disappeared under the use of common applications (Bennett 1845, referring to the work of Professor Vogel of Munich).

Within the last year, we ourselves have been able to diagnose cancer in the living subject, in six instances, by nipping off little projecting points of the ulcerated surfaces. In one case, at the request of Professor N. R. Smith, of a patient at the Baltimore Infirmary, when there was extensive disease, with induration and ulceration of a doubtful character, of the penis; in another, by the kindness of my friend, Dr. Van Bibber, in a patient of his, suffering with a tumour accompanied with deep-seated ulceration of the posterior fauces; in a third, a patient of Dr. Maris's, there was a large encephaloid tumour of the neck; in two cases of disease of the neck of the uterus; and in another ulcerated penis. In all, the microscope revealed unmistakable evidences of cancer (Donaldson 1853).

Credit must likewise go to Lionel Beale, Professor of Medicine at King's College Hospital, London, whose pioneering treatise *The Microscope in Medicine* (1854) firmly established the microscope as a diagnostic tool of considerable versatility. The microscopist could be found "continually fishing for new facts in the excretions of the sick". Much of this early work of Bennett, Donaldson, Beale and others was done with effusions, exudates, smears and tissues teased in saline and consequently cytology was regarded as the "senior science" for many years (Bamforth and Osborn 1958). Only later were sections cut on a regular basis for diagnostic purposes.

The enthusiastic reception that greeted biopsy diagnosis was, in Europe at least, sharply punctuated by the case of the Kaiser's cancer. In 1887, Kaiser Frederick III had a laryngeal lesion biopsied by Sir Morell Mackenzie. The biopsies were reported as benign by Rudolf Virchow in Berlin and to judge from his published descriptions, the actual tissue submitted to him may well have been free from malignancy (Virchow 1887). (As so often happens with interesting cases, the blocks and slides are no longer available for review.) One year later, at the age of 56, the Kaiser died from laryngeal cancer. This infamous case ruined Mackenzie's reputation and started a wave of opposition against the technique of biopsy diagnosis. Virchow, naturally a conservative interpreter of biopsies, was at the forefront of the opposition (Ober 1970).

In 1879 Carl Ruge, a gynaecologist at the Women's Hospital in Berlin, reported that the microscopic examination of curettings could be used as a method for the diagnosis of cancers of the uterine cervix; 2 years later he advocated the same method for the diagnosis of endometrial cancer. Many German pathologists protested that cancer could be diagnosed only by observing invasion of deeper structures and the presence of metastases at autopsy. Perhaps their more cautious attitude stemmed from Virchow's widely publicised misdiagnosis of the Kaiser's laryngeal cancer as a benign lesion. Ruge's attitude was that clinicians should perform and examine their own biopsies and bypass the academic pathologists who were unprepared to venture into this new era of diagnosis.

Unabated interest in biopsy diagnosis in North America was marked, in the 1890s, by reports from W. H. Welch and T. H. Cullen on rapid frozen section diagnosis during surgical operations. From 1900 onwards, the biopsy was increasingly accepted throughout the medical world as a significant advance. The Kaiser's cancer was quickly forgotten.

The first appearance in hospitals of pathology, as we would recognise it today, evolved from the collecting of museum specimens primarily for educational purposes. This was usually done by physicians and surgeons, many of whom became very skilled at the recognition of gross morbid changes. This practice preceded by some years the identification of gross and microscopic abnormalities in tissues and body fluids as an adjunct to clinical diagnosis.

Rapid advances in pathology ultimately created insuperable difficulties for physicians and surgeons doing histopathology and cytology. The discipline had increased in complexity as more and more knowledge was gleaned from the application of new methods to pathological material. The sheer volume of diagnostic work was also increasing and this necessitated, in most centres, the appointment of a specialist in general pathology who would have responsibility for haematology, bacteriology and clinical chemistry, as well as for histopathology. Continuing advances in all four subspecialties made unacceptable demands on the general pathologist forced to keep abreast with laboratory work of such diversity. Thus emerged the specialist histopathologists and surgical pathologists.

The American College of Surgeons played a major role in establishing the requirement for properly staffed laboratories in hospitals in the U.S.A. (Wright 1985). The minimum standards recommended by the College in 1918 were that hospitals should provide "chemical, bacteriological, serological, and pathological services . . . under

competent medical supervision" (cited by Wright 1985). In 1926 the College's requirement became even more stringent: "the clinical laboratory shall be under the direction
of a graduate in medicine, especially trained in clinical pathology" and "all tissues
removed at operation shall be examined in the laboratory and reports rendered
thereon" (cited by Wright 1985). These requirements reflected the growing importance of surgical pathology and the increasing dependence of clinical practice on proper
histopathological support, and paralleled similar trends in many other countries.

Diagnostic cytology proceeded along similar lines. Papanicolaou's interest in the
cytology of the female genital tract led, in 1928, to the first report of his studies on
cervical cancer at the Women's Hospital of New York City (Papanicolaou 1928).
By the 1950s cervical cytology was used as a screening test for the early detection
of this disease. Cytology as an aid to the diagnosis of cancer in other organs developed
similarly (Bamforth and Osborn 1958).

A chronology is given in Table 1.1.

Table 1.1. Historical summary of the development of diagnostic histopathology

Pre-1800	Descriptive morbid anatomy (Galen, Severino, Morgagni, Baillie et al.)
1800–1840	First descriptions of tumour histology (Home, Müller et al.)
1840–1855	Advent of diagnostic cytology and histopathology (Bennett 1845; Marmy 1846; Lebert 1851; Donaldson 1853).
ca. 1880	Paraffin wax embedding
1887	Biopsy, reported by Virchow, fails to diagnose cancer in Kaiser Frederick III
1891	First intra-operative frozen section (Welch)
1928	Diagnostic cytology in cervical cancer (Papanicolaou)
1930–1940	Introduction of needle biopsy techniques
1938	Linderstrøm-Lang cryostat for histochemistry
1941	Fluorescent-labelled antibody technique (Coons et al.)
1943	Application of stereological methods to tissues (Chalkley)
ca. 1950	Electron microscopy applied to tissues
1961	Cryostat used for intra-operative diagnosis (Chang et al.)
1966	Introduction of enzyme (e.g. peroxidase)—antibody conjugates (Nakane and Pierce)
1970	Application of microspectrophotometry and flow cytometry
	Mathematical treatment (cluster, vector analysis) of diagnostic problems
	Evaluation of diagnostic fallibility in histopathology
1980-	Gene probing (in situ hybridization)

The Objectives of Histopathology

Clinical Diagnosis

The routine performance of diagnostic histopathology by clinical specialists who have
primary responsibilities for patient care is considered by many to be potentially
hazardous. The assessment of biopsies objectively and unbiased, in the first instance,
by detailed knowledge of the full clinical picture has many advantages. It is easy
from the standpoint of a strong clinical conviction to read into a biopsy an interpretation that may be quite unsupported by objective evidence. It is impossible for those
who are familiar with a particular patient to be absolutely impartial. Ultimately the
pathologists may become familiar with the clinical picture, but the patient will at
least have had the benefit of another opinion arrived at independently. There is, how-

ever, considerable merit in joint discussions with other specialists at which the biopsy is presented and discussed. The formal participation of histopathologists in clinical teams is now commonly advocated for the management of patients with malignant lymphomas, liver disease and renal disease among other entities.

Exacting demands are made on the modern histopathologist. Advances in endoscopy and needle biopsy techniques have led to involvement in the diagnosis and management of diseases previously beyond the reach of biopsy; we are now called upon to make more precise diagnoses on smaller and smaller fragments of tissue. Immunology, biochemistry and physics have contributed an array of new investigative methods, each requiring further interpretative skills to assess the results. Tumours previously dismissed as anaplastic, or perhaps because of arbitrariness or marginal features tentatively placed in a particular category, must now go through the gamut of electron microscopy and immunohistochemistry for marker organelles or substances. Advances in chemotherapy, radiotherapy and endocrine therapy exactly tailored to specific tumour types must be paralleled by refinements in biopsy diagnosis and classification.

The proliferation of needle and endoscopic biopsy techniques has led to a more detailed understanding of many disease processes and has greatly enhanced histopathological contributions to the care of patients with disorders amenable to both medical and surgical treatment. Biopsies of many organs can now be obtained, though a compromise has to be found between the volume of tissue required for reliable interpretation and that which can be removed from the patient with acceptable safety. Any risks to the patient must be outweighed by the possible therapeutic benefit.

In most hospitals the histopathologist is also responsible for the autopsy service. Although this is traditional, it reflects the continuing interplay between post-mortem findings and the recognition at autopsy of diagnostically useful information that may benefit future patients. The relative clinical independence of histopathologists favours their role as convenors of "Tissue Committees", which investigate allegedly superfluous surgery, and "Death Committees", which police hospital mortality. These audit committees are more popular in North America than elsewhere.

In essence, therefore, the primary clinical role of the diagnostic histopathologist is to provide a reliable and efficient diagnostic service through the interpretation of biopsies and the assessment of surgical resections. Close liaison with clinical teams is vital to the management of many patients.

Epidemiology

Histopathologists are ideally placed to study the epidemiology of disease for two reasons. First, they see a large volume of material by way of biopsies and autopsies. Second, precise classification and uniformly consistent nomenclature in any one district makes it easier to collect and investigate patients with the same disorder. In a continually changing environment, all medical workers need to be alert to the emergence of novel disease entities or unexpected changes in disease patterns and incidence. With respect to carcinogens, this subject has recently been reviewed by Higginson (1977). Carcinogenic agents are not generally visible in tissues, though there are exceptions. Asbestos may be seen in lung adjacent to a mesothelioma or pulmonary carcinoma, thorotrast in an hepatic angiosarcoma, *Clonorchis sinensis* and *Schistosoma* spp. in cholangiocarcinoma and bladder carcinoma respectively.

Histopathologists are also called upon to identify and determine the significance of precancerous or other pathogenetically associated lesions (Morson 1979). Screening of an asymptomatic population or groups at risk is a major task for the cytologist. It is obviously less traumatic to scrape off some cells from a body surface than it is to remove a solid lump of tissue. The volume of work is enormous and usually calls for a team of "screeners" to select the few smears that need more careful scrutiny. Advances in electronic image analysis and cell sorting may eventually lead to the routine use of automatic devices for population screening.

Treatment Selection and Monitoring

Biopsies are often done not only to establish the diagnosis but, in addition, to determine the extent of the lesion so that treatment can be appropriately chosen to fit the disease. This is well illustrated by the use of laparotomies in patients with Hodgkin's disease followed by histological examination of lymph nodes, liver biopsies, and splenectomies to confirm or exclude abdominal involvement, though this strategy is now less common.

Responses to treatment may be followed in biopsies and may support or refute clinical assessment of the efficacy of therapy. A common example is the identification of villous regrowth and a reduction in the intraepithelial lymphocyte population in the jejunal biopsy of a patient with coeliac disease treated by gluten-free diet. Tumour regression induced by treatment may be monitored by biopsy. Without biopsy, for example, it may clinically be difficult to distinguish a radionecrotic lesion from residual or recurrent tumour. Even in cases where neoplastic tissue is still evident in post-treatment biopsies, its integrity is sometimes a useful guide to its therapeutic sensitivity.

Attempts to predict the chemotherapeutic sensitivity of tumours have met with variable success. Most of these studies have employed short-term tissue or organ cultures to which a range of agents is added. The structural or histochemical integrity of the tissue is assessed at the end of the assay. The hormone dependence of breast carcinomas has been determined in a similar way, or in the case of oestrogen and progestogen receptors by the ability of the soluble fraction of a tumour homogenate to bind radiolabelled hormone.

Equally vital is the identification of the side effects of treatment. The identification of total novel toxic effects, such as sclerosing peritonitis following practolol therapy, must be credited to the open and inquiring minds of those who first encounter these conditions.

Secretory products of tumours can be identified in biopsies by the application of immunohistochemical methods. Where no pre-operative serum samples are available, the recognition of a specific secretory product in tumour tissue can be a fairly reliable indication that the appearance of the product in serum post-operatively will reflect residual or recurrent disease.

Clinical Research

Medicine is a dynamic subject. During the nineteenth century, many advances in medical knowledge directly resulted from the application of histopathological methods to disease. To a modest extent the approach is still active and, no doubt, will continue.

There is now a trend, however, for histopathology to apply knowledge gained from other disciplines to develop improved diagnostic methods. This can be seen, for example, by the application of radiology, electron optics, molecular biology and immunology to the study and diagnosis of disease in histopathology laboratories. Histopathology must be flexible and adapt to changing circumstances (Kornberg 1977), rather than restrict itself to purely traditional methodology.

With respect to specific clinical research projects, it is useful if the project is discussed with an interested histopathologist beforehand if it is contemplated that biopsy analysis should constitute part of the protocol. The proposed biopsy technique must be one that will yield a sample of acceptable quantity and quality. Biopsies should be assessed blindly to avoid bias; if two or more treatment groups are to be compared, the biopsies must not be identified as to their group of origin until an assessment has been made.

The responsibilities of histopathologists for the conduct and facilitation of research are the subject of recent reviews (DeLellis 1981; Silverberg 1981).

Autopsies and Clinicopathological Correlation

The autopsy has a continuing role in establishing diagnostic criteria against which biopsies are interpreted. The individual histopathologist, particularly the trainee, gains potentially valuable diagnostic information from a careful analysis of gross and microscopic autopsy appearances. Various organ biopsies containing tumour too poorly differentiated to betray any specific origin are often seen; autopsies on such patients are important to establish, if possible, the primary site both for epidemiological purposes and for more accurate assessment of similar biopsies in future.

Autopsies are also valuable even when the diagnosis, made on a previous biopsy, is not apparently in doubt. For example, needle biopsies of liver from patients with chronic inflammatory liver disease and cirrhosis can be compared with autopsy appearances, gross and microscopic, to provide an insight into the degree of sampling error. The dynamics of a disease can be appreciated by comparison of a biopsy with the subsequent autopsy findings. The autopsy still has considerable educational value for pathologists and clinicians alike. Even in "clear-cut cases" approximately 50% of contributory disorders may be unrecognised before death (Britton 1974a,b).

Clinicopathological correlation is therefore essential for the conduct of high-quality diagnostic work. It is increasingly common for the histological appearances of diseases to be modified by therapy and novel or puzzling features may be resolved only after full appraisal of the clinical circumstances.

Medicolegal Aspects

The medicolegal aspects of surgical pathology are to some extent dependent on national or local laws and regulations, but the principles of patient confidentiality and the general ethics of pathological practice are universal and apply just as much to pathology as they do to any other branch of medicine. Some specific medicolegal aspects of histopathology are considered here. Medicolegal aspects of the autopsy are considered in Chap. 13.

Histopathologists do not commonly have patient contact, but the indirectness of their relationship does not exempt them from litigation instigated by patients. Even if the conduct of the histopathologist and the laboratory is not the matter at issue, it is nevertheless essential that all the procedures and circumstances surrounding the event which is the subject of litigation should be adequately documented and that includes the logging of specimens, their gross description, and their histological interpretation. The person signing the final report, even though he or she may not have been directly responsible for all stages of the handling of that particular specimen is, in so doing, endorsing not only the diagnosis but all the statements in that report. If an allegedly erroneous histological diagnosis is at issue (e.g. an allegedly unnecessary mastectomy performed for a mistaken diagnosis of carcinoma on a frozen section) then it is absolutely vital that the documentation of the diagnosis and the events surrounding it are beyond reproach.

The length of time that specimens are kept will vary from one department to another in the absence of national guidelines or regulations. The Colorado Society of Clinical Pathologists for example suggests a minimum retention period of 1 month for wet tissue, 2 years for paraffin wax blocks and accession logs, 5 years for reports, and 20 years for slides (Hoffman and Silverberg 1983). Most departments will be able to retain material for much longer than this, and now that immunohistology and even electron microscopy can be done on archival blocks most histopathologists will be reluctant to throw anything away. In the United Kingdom there is a legal requirement to retain all patient-related material for a minimum of 8 years after which it can be discarded.

In resolving a difficult differential diagnosis between benign and malignant, the pathologist must avoid the temptation to call the lesion "malignant" on the assumption that it is better to overdiagnose a benign lesion as malignant than to risk missing a potentially fatal malignant lesion. A "safe" diagnosis of malignancy on an equivocal but benign lesion may result in further unnecessary surgery leading to avoidable disfigurement, disability, complications or even mortality. It may be only when the original histology is reviewed in the light of subsequent events that the diagnosis is questioned thus exposing the pathologist to the threat of litigation (Grossman 1981). One must, however, distinguish between errors of judgement which, at least in English law, are not culpable and mishaps arising from negligence or incompetence which may well prove to be indefensible. For a mishap to be attributable to an error of judgement it must be demonstrated that the histopathologist was exercising the same general levels of skill, competence and knowledge that would be exercised by other histopathologists of similar experience working under the same conditions. Failure to apply the same general levels of skill, competence and knowledge either by omission or commission expose the histopathologist to the risk of successful litigation.

The Histopathological Diagnosis

The "tissue diagnosis" holds a revered place in clinical medicine. To it is awarded a degree of credence and infallibility that is rarely questioned.

Diagnosis and Disease

Making a diagnosis is the act of recognising a disease, or distinguishing it from another, and assigning a name to it. Before we consider the ways in which a histopathological diagnosis is made, and its validity, we must consider the nature of diseases and diagnoses. This is the subject of scholarly reviews by King (1967a) and Bohrod (1971).

Diseases are recognised from disturbances in the structure and function of tissues and their constituent parts. Diseases do not *cause* these disturbances; the disturbances are the hallmarks of the diseases. Recognition of disease is done, therefore, by comparing the patient or lesion in question with what we would regard as the normal state. We must accept that normality is not a discrete single locus but rather the bell-shaped curve of the normal distribution. Where quantitation is possible, such as with height, weight, blood pressure and so on, the normal distribution has a measurable reality that can easily be grasped and it enables us to set, albeit arbitrarily, defined limits to the normal range. With histopathology, quantitation is often either impossible or too cumbersome to be routinely applied; the background of normality against which biopsies are compared is derived largely from subjective experience, and must encompass the natural variations seen for example in relation to race, gender, puberty, pregnancy and age.

Having perceived the presence of disease, it then has to be assigned a name by recognising in the biopsy certain features which are common to that named disease. Through a process of "lumping" or "splitting" (see p. 181) we have arrived at a series of classifications, admittedly subject to change, in which diseases form specific units or categories. As Bohrod emphasises, these classifications are arbitary inventions, but not haphazard. They are designed to fulfil certain purposes; if they fail to do so, they must be reconstructed.

Classifications, then, are not right or wrong; they cannot even be said to be good or bad except in relation to a purpose. The most that can be said about them is that they are useful or not useful.... It was as accurate and factual for our forebears to classify whales with fishes as for us to find this classification abhorrent (Bohrod 1971).

Classifications of disease can themselves be grouped according to their methodological origin—histopathological (e.g. membranous glomerulonephritis), immunological (e.g. immune-complex nephritis), clinical (e.g. nephrotic syndrome) and so on. Different classifications may show high degrees of correlation with each other; this can be useful, though it is not an absolute prerequisite for success. Other classifications include aetiological (e.g. post-streptococcal glomerulonephritis), and therapeutic. Rademacher (1772–1850) (cited by Bohrod 1971) considered that diseases should be named according to the treatment that was most successful (e.g. in contemporary practice—steroid-responsive nephropathy).

The role played by histopathology in diagnosis varies greatly from one disease to another. Some diseases are diagnosed without resorting to biopsy at all, such as ischaemic heart disease. Others involve a clinico-radiological approach, such as secondary hypertension due to renal artery stenosis. Yet others can only be diagnosed by biopsy if the histological appearances are seen in the right clinical context. In only a few diseases is a pathognomonic pattern imprinted on such small pieces of tissue that, alone, the histopathologist can make the diagnosis; this is the case with most tumours.

Ultimately the name agreed upon for a disease may be eponymous, generally after the person credited with its original description, though this can be controversial if the claim to originality is disputed. While some fight against eponymous names, those of the stature of Addison, Cushing and Paget, among many others, are indelibly etched into the language of medicine. Crohn's disease is the best name for a condition which was originally described as terminal ileitis, reappearing as regional ileitis, because some were misled into thinking that "terminal" was temporal (i.e. agonal) rather than anatomical, then as regional enteritis when it was found outside the ileum. Now that we recognise it from the mouth to the anus, an eponymous designation eliminates the need for any further changes.

With this understanding of what constitutes a diagnosis and a named disease, we can turn our attention to the actual diagnostic process itself.

Diagnostic Processes

King (1967a) states that there are two steps involved in making a diagnosis— knowledge and judgment. Knowledge is the acquisition of information about various diseases. This can be easily taught and put into textbooks. It includes descriptions of structural changes in diseased tissues. Judgement is the skill, intimately linked to experience, which enables a pathologist to recognise and categorise lesions. One pathologist may be very knowledgeable about a particular lesion but, lacking practical experience, may fail to recognise it. The good diagnostician not only has the knowledge about a disease, but is capable of reliably recognising it.

Feinstein (1967) claims that pathologists make diagnoses simply by looking, without exercising inference. King (1967b), on the other hand, has argued strongly against this proposition. King's contention that the clinician and the pathologist use the same intellectual processes to make a diagnosis is one with which most histopathologists would agree. Both must accurately record what they perceive in a patient or biopsy, and infer from what they perceive to make a diagnosis. Judgement and intuition may be necessary to resolve a differential diagnosis. Feinstein concedes that pathologists engage in inferential or deductive analysis only when speculating about the cause of a disease. This attitude may arise from the common practice of merely stating, on the biopsy report, the pathological diagnosis without actually describing the appearances from which the diagnosis was inferred. It is usually superfluous to give a detailed description of a straightforward simple pigmented naevus or lipoma; it is just necessary to state what the lesion is, as a matter of opinion, and comment on the adequacy of excision. In contrast, the reactive or neoplastic lymph node and the liver biopsy, among other examples, demand a detailed description of the appearances in addition to the inference with which the report should end.

The fallibility of histopathological diagnoses is an inevitable consequence of the existence of a major component of intuition and judgement in biopsy interpretation.

Diagnostic Fallibility

Contrary to the popular belief that the "tissue diagnosis" has a veracity that cannot be questioned, it shares with many other diagnostic procedures an appreciable measure of fallibility (Desjardins 1960).

Distinction must first be drawn between the fallibility, credibility, and plausibility of a histological diagnosis. Fallibility is an expression of error. Credibility is the extent to which a diagnostic opinion can be trusted or believed. Plausibility does not question the validity of an opinion, merely whether the biopsy diagnosis is consistent with the clinical picture. Consider, for example, a bone biopsy from a site at which osteosarcoma is suspected on clinical and radiological grounds. Let us assume for these purposes that the biopsy unquestionably shows fracture callus, an opinion reached unanimously by a panel of internationally acclaimed doyens. The biopsy may be wrongly reported as osteosarcoma because, despite having seen the radiological evidence, having heard the clinical story, and, having been aware of the way in which callus can mimic sarcoma, the pathologist made an error of judgement. This error indicates fallibility. A second pathologist may correctly report the biopsy as showing fracture callus rather than sarcoma. However, this pathologist may be well known in orthopaedic circles to be prone to erroneous diagnoses and to indulge in the deplorable habit of expressing an opinion on bone biopsies without seeing X-rays of the lesion. Both the report and the pathologist lack credibility. Little trust can be placed in the report, despite its accuracy on this occasion. A third pathologist may, having seen the X-rays and looked into the clinical story, report the biopsy as showing only fracture callus. This is a valid interpretation but lacks plausibility because it conflicts with the radiological and clinical evidence. The report can only acquire plausibility if it includes an admission that the biopsy appearances do not support the clinical diagnosis and that this may be due to an inadequate sample from the edge of the lesion.

Diagnostic errors are not peculiar to histopathology. They are well documented in the interpretation of radiographs, electro-cardiographs, and clinical assessments of various disorders (Garland 1959). Most of us are aware of the existence of diagnostic mistakes, both our own and those of our colleagues.

Studies on specific groups of lesions have given some indication of the extent of unreliability in histological interpretation.

1. *Lymphomas.* Observer inconsistencies in the classification of Hodgkin's lymphoma led Coppleson et al. (1970) to conclude that evaluation of such biopsies by individual observers was unacceptable for scientific use. They recommended a consensus by a team of pathologists. A similar study was conducted on biopsies of non-Hodgkin's lymphoma (Iversen and Sandnes 1971).

Symmers (1968) found that in a series of 600 lymph node biopsies initially diagnosed as Hodgkin's disease, the diagnosis was confirmed on review in a reference laboratory in only 53% of cases. The author included Hodgkin's disease in a short list of his own "personal diagnostic quicksands".

2. *Lung Tumours.* Five pathologists participated in an elegant study of the authenticity and reproducibility of lung tumour classifications (Feinstein et al. 1970). Each pathologist was given one section from each of 50 lung tumours, numbered from 1 to 50, and asked to describe and interpret the appearances free from the constraints of any particular classification. They were told to expect sections from a second set of 50 tumours when they had completed and returned their assessment of the first set; they were not told that this second set of slides was simply the first set renumbered from a table of random numbers. The results could therefore be analysed for inter-observer and intra-observer variability. In summary, well-differentiated tumours led to few inconsistencies, whereas the disagreement rate with poorly differentiated

tumours was as high as 40%. Intra-observer inconsistency ranged from 2% for the most consistent to 20% for the least consistent.

3. *Liver Biopsies.* The diagnostic consistency of three pathologists was tested by challenging them with 51 liver biopsies (Garceau 1964). They arrived at identical diagnoses in only 21 instances and offered a total of three different diagnoses on 13 of the biopsies. Paradoxically, they were unanimous in their assessment in 52% of the 29 needle biopsies but agreed on only 27% of the larger wedge biopsies. The reasons for this anomalous result are uncertain. The larger sample should theoretically provide a larger quantity of evidence on which to make a judgement, but fewer observations can be made on small biopsies and this might in turn raise fewer diagnostic possibilities. Soloway et al. (1971) found that sampling was probably a greater source of error than observer variability, particularly in cirrhosis. Theodossi et al. (1977) in a similar study found 80% agreement between two experienced histopathologists, but the consensus rate dropped to 56.7% when a trainee was included in the panel.

I know of no attempts to elucidate those factors which cause variations in the interpretation of a single sample by the same observer. No doubt ambient conditions play a part; adverse circumstances may sway interpretation of a borderline lesion towards malignancy or a liver biopsy towards chronic active hepatitis. Sissons (1975), commenting on observer disagreement in bone tumour interpretation, remarks upon the paucity of work done in this area.

How can observer variability and diagnostic fallibility be reduced? Diagnoses made by consensus among a group of pathologists tend to be more reliable and reproducible than those made by individuals; this approach is usually impracticable for routine use, but can be useful in selected cases. Some have advocated more widespread application of novel technology, such as histochemistry and electron microscopy (Pearse 1975). These techniques certainly provide additional information to help resolve a diagnostic problem, as do special stains, but the final step is still one of subjective interpretation. Others seek a solution in greater objectivity achieved through the

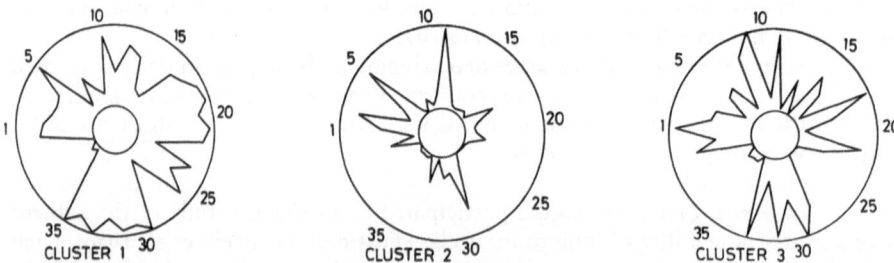

Fig. 1.1. Polar vector diagrams of the histological characteristics of computer-formed clusters (groups of cases) or oral mucosal lesions; cluster 1 depicts carcinoma and "severe" leukoplakia (20 cases), cluster 2 shows mainly benign keratosis (102 cases), and cluster 3 shows lichen planus (34 cases). For every biopsy, 2 clinical and 39 histological variables were scored either 1 (present) or 0 (absent) and the mean values in each cluster were graphically plotted along 41 equally spaced radii; maximum values reach the circumference. Discriminating variables include 1 (multiple lesions), 5 (acanthosis), 9 (liquefaction degeneration of basal layer), 24 and 25 (increased mitotic activity in stratum spinosum and basal cell layer respectively) and 35 (Russell bodies in lamina propria). This type of analysis can be used to determine discriminating variables objectively as an aid to more accurate diagnosis. (Kramer et al. 1974)

application of logical and mathematical analysis (Lusted and Ledley 1960; Kramer 1975; Zajicek et al. 1977). Histopathologists generally interpret a specimen by the subjective assessment of one or more perceived abnormalities and subconsciously compare that assessment with a memory store of knowledge and experience. However, the raw data obtained from a gross specimen of the microscope image is incredibly complex and Kramer suggests that objectivity can be improved by recording specific features according to a set of standardised criteria. This information can then be analysed statistically and some degree of significance attached to the result. For example, the malignant potential of oral leukoplakia has been assessed using polar vector diagrams and cluster analysis (Fig. 1.1; Kramer et al. 1974).

It is likely that the continuing interest in computer diagnosis will necessitate a more widespread adoption of standard histological criteria against which biopsies can be objectively assessed. Direct analysis of histological images by computer has been attempted, but the diversity of morphological patterns prohibits universal application (Bartels et al. 1976).

Adverse Effects of Biopsy

Biopsies of any type are invasive procedures and, however carefully they are performed, there is a variable risk of consequent morbidity and mortality. Some biopsies (e.g. renal and liver) are more hazardous than others (e.g. rectal and skin), but the relative risk and the possible therapeutic benefit shold be considered in every case. Specific risks of biopsies include:

1. Vascular problems such as: bleeding from the biopsy site evident either by external haemorrhage, internal haemorrhage into a body cavity, hollow viscus, or duct (e.g. haemobilia following a percutaneous liver biopsy); or the formation of an arterio-venous fistula after needle biopsy of a vascular organ (e.g. kidneys)

2. Perforation of a viscus, duct, or organ resulting in leakage of contents into another structure (e.g. biliary peritonitis after liver biopsy and pneumothorax after a lung biopsy)

3. Dissemination of tumour or seeding of the biopsy tract (an exceptionally rare event)

4. Induction of reactive changes which on subsequent biopsy may be misinterpreted, e.g. reactive changes or infarction in lymph nodes (Davies et al. 1982; Steele et al. 1983); prostatic granulomas after previous biopsy or instrumentation (Lee and Shepherd, 1983); pseudomalignant appearances in moles previously subjected to shave biopsy (Ackerman 1983)

5. Infection either from a contaminated biopsy instrument or from a biopsy site serving as a portal of entry for micro-organisms

Biopsies should therefore never be performed out of idle curiosity; there must be a real clinical indication and preferably a genuine possibility of therapeutic benefit to the patient based on the interpretation of the biopsy. Any biopsy procedure outside these limits constitutes research and requires ethical approval and informed consent from the patient.

Prospects for Histopathology

Advances in histopathology arise from the application of new methods, exemplified by the enormous impact of immunological methods and electron microscopy during the last three decades, and from the extended anatomical range of biopsy procedures, particularly endoscopy. Occasionally the advance is activated not by the innate curiosity of histopathologists, but is forced by advances in therapy that demand greater accuracy and precision in the biopsy diagnoses. In some situations, the efforts made by the histopathologist to reach a very precise diagnosis, such as assigning a non-Hodgkin's lymphoma to one of the many categories in the plethora of current classifications, far exceed the clinician's ability to prescribe a therapy that has commensurate specificity; in many centres the classification of a non-Hodgkin's lymphoma into either a low or high-grade category is sufficient for therapeutic purposes. Nevertheless, this should not deter histopathologists from advancing their own subject in anticipation of advances in therapy which could take advantage of more accurate and precise biopsy diagnoses.

The extended anatomical range of biopsy procedures has resulted in an increased workload for most laboratories, particularly from endoscopies. A survey of endoscopists in the United Kingdom revealed that the number of diagnostic procedures between 1972 and 1978 had increased dramatically: oesophagus, stomach and duodenum—from 19 000 to 110 000; retrograde cholangio-pancreatography—from 450 to 4450; and colonoscopy—from 1000 to 6200 (Cockel et al. 1982). Not all of these endoscopies generate biopsies, but most laboratories will have experienced a marked increase in their workload from these sources during this period, and there is every indication that this will continue. Similarly, advances in organ imaging such as ultrasound and computerized axial tomography have resulted in an increased number of biopsies or cytology specimens from lesions that were previously difficult to target (e.g. retroperitoneal tumours). To maximise the diagnostic usefulness of these developments in imaging and histopathology, close collaboration between histopathologists and the other specialists involved is essential.

It is likely that the revolution in immunohistology will continue with an increasing repertoire of exquisitely specific monoclonal antibodies. Hopefully there will be further developments akin to the immunogold-silver technique which will enable the demonstration of antigens adversely affected by processing tissue into paraffin wax blocks, such as most of the lymphocyte surface markers.

Cytology is likely to increase in importance because: it is less traumatic and more acceptable to patients; diagnostic criteria are undergoing continuing improvement; diagnostic precision will be improved by the use of immunological methods and electron microscopy; developments in image analysis may result in the introduction of automatic systems; and advances in imaging referred to above enable previously inaccessible lesions to be sampled.

There is currently much interest in the application of gene analysis to diagnostic problems in histopathology particularly in the field of tumour pathology, but the resources are available only in very few institutions. Specific rearrangement of immunoglobulin and T-cell receptor genes are now being used to type lymphomas (O'Connor et al. 1985), possibly eventually displacing the immunophenotype as the "gold standard" for lymphoma diagnosis.

References

Ackerman AB (1983) Shave biopsies: the good and right, the bad and wrong. Am J Dermatopathol 5: 211–212

Bamforth J, Osborn GR (1958) Diagnosis from cells. J Clin Pathol 11: 473–482

Bartels P, Bibbo M, Wied G (1976) Modeling of histologic images by computer. Acta Cytol (Baltimore) 20: 62–67

Beale LS (1854) The microscope in medicine. Churchill, London

Bennett JH (1845) Introductory address to a course of lectures on histology and the use of the microscope. Lancet I: 517–522

Bohrod MG (1971) What is a pathologic diagnosis? Pathol Annu 6: 197–208

Britton M (1974a) Diagnostic errors discovered at autopsy. Acta Med Scand 196: 203–210

Britton M (1974b) Clinical diagnostics: experience from 383 autopsied cases. Acta Med Scand 196: 211–219

Cockel R, Colin-Jones DG, Schiller KFR (1982) Gastro-intestinal endoscopy services: a review of the 70s with predictions for the 80s. Health Trends 14: 46–49

Coppleson LW, Factor RM, Strum SB, Graff PW, Rappaport H (1970) Observer disagreement in the classification and histology of Hodgkin's disease. J Natl Cancer Inst 45: 731–740

Davies JD, Webb AJ (1982) Segmental lymph-node infarction after fine-needle aspiration. J Clin Pathol 35: 855–857

DeLellis RA (1981) The future of research in surgical pathology. Am J Clin Pathol 75 [Suppl]: 476–482

Desjardins AU (1960) Is the pathologist infallible? Arch Intern Med 106: 596–602

Donaldson F (1853) The practical application of the microscope to the diagnosis of cancer. Am J Med Sci 25: 43–70

Feinstein AR (1967) Clinical judgement. Williams and Wilkins, Baltimore, p 80

Feinstein AR, Gelfman NA, Yesner R (1970) Observer variability in the histopathologic diagnosis of lung cancer. Am Rev Respir Dis 101: 671–684

Garceau AJ (1964) The natural history of cirrhosis. II. The influence of alcohol and prior hepatitis on pathology and prognosis. N Engl J Med 271: 1173–1179

Garland LH (1959) Studies on the accuracy of diagnostic procedures. AJR 82: 25–38

Grossman SZ (1981) Legal implications of overdiagnosing malignant melanoma. Am J Dermatopathol 3: 67

Higginson J (1977) The role of the pathologist in environmental medicine and public health. Am J Pathol 86: 460–484

Hoffman DS, Silverberg SG (1983) Medicolegal principles and problems. In: Silverberg SG (ed) Principles and practice of surgical pathology. Wiley, New York, pp 21–25

Iversen OH, Sandnes K (1971) The reliability of pathologists. A study of some cases of lymph node biopsies showing giant follicular hyperplasia or lymphoma. Acta Pathol Microbiol Immunol Scand [A] 79: 330–334

King LS (1967a) What is a diagnosis? JAMA 202: 154–157

King LS (1967b) How does a pathologist make a diagnosis? Arch Pathol 84: 331–333

King LS, Meehan MC (1973) A history of the autopsy: a review. Am J Pathol 73: 514–544

Kornberg A (1977) Pathology, pathologists and the new biology. Arch Pathol Lab Med 101: 397–399

Kramer IRH (1975) Computer-aided analyses in diagnostic histopathology. Postgrad Med J 51: 690–694

Kramer IRH, El-Labban NG, Sonkodi S (1974) Further studies on lesions of the oral mucosa using computer-aided analysis of histological features. Br J Cancer 29: 223–231

Lee G, Shepherd N (1983) Necrotizing granulomata in prostatic resection specimens: a sequel to previous operation. J Clin Pathol 36: 1067–1070

Lusted LB, Ledley RS (1960) Mathematical models in medical diagnosis. J Med Educ 35: 214–222

Morson BC (1979) Prevention of colorectal cancer. Proc R Soc Med 27: 83–85

Ober WB (1970) The case of the Kaiser's cancer. Pathol Annu 5: 207–216

O'Connor NTJ, Wainscoat JS, Weatherall DJ et al. (1985) Rearrangement of the T-cell-receptor β-chain gene in the diagnosis of lymphoproliferative disorders. Lancet I: 1295–1302

Papanicolaou GN (1928) New cancer diagnosis. In: Proceedings of the 3rd Race Betterment Conference. Race Betterment Foundation, Battle Creek, Michigan, pp 528–534

Pearse AGE (1975) The role of histochemistry in increasing objectivity in histopathology. Postgrad Med J 51: 708–710

Silverberg SG (1981) The surgical pathologist as researcher. Am J Clin Pathol 75 [Suppl]: 453–456

Sissons HA (1975) Agreement and disagreement between pathologists in histological diagnosis. Postgrad Med J 51: 685–689

Soloway RD, Bagenstoss AH, Schoenfield LJ, Summerskill WHJ (1971) Observer error and sampling variability in evaluation of hepatitis and cirrhosis by liver biopsy. Am J Dig Dis 16: 1082–1086

Steele RJC, Blackie RAS, McGregor JD, Forrest APM (1983) The effect of breast biopsy on reactive changes in axillary lymph nodes. Br J Surg 70: 317–318

Symmers WStC (1968) Survey of the eventual diagnosis in 600 cases referred for a second histological opinion after an initial biopsy diagnosis of Hodgkin's disease. J Clin Pathol 21: 650–653

Theodossi A, Knill-Jones RP, Silk DBA, Williams R (1977) Study of interobserver variation between liver histopathologists. Gut 18: 953

Virchow R (1887) Professor Virchow's report on the portion of growth removed from the larynx of H.I.H. the Crown Prince of Germany by Dr. M. Mackenzie on June 28th. Br Med J II: 199

Wright JR (1985) The development of the frozen section technique, the evolution of surgical biopsy, and the origins of surgical pathology. Bull Med Hist 59: 295–326

Zajicek G, Maayan C, Rosenmann E (1977) The application of cluster analysis to glomerular histopathology. Comput Biomed Res 10: 471–481

2 Macroscopy, Microscopy and Sampling

The gross examination and sampling of surgical resections and biopsy specimens for histology is an important responsibility of the histopathologist, detailed practical guides to which can be found elsewhere (Kennedy 1977; Pierson 1980; Rosai 1981). Blocks of tissue for histology should always be taken with due regard to the particular problem under investigation, conscious of the local anatomy, and with the aim of minimising sampling error.

Sampling Error

An inherent risk in the examination of any object of non-uniform composition is that a small portion may not be adequately representative of the whole. Sampling error describes the difference in composition between the whole object and that portion selected for examination. The magnitude of the error is inversely proportional to the relative sample size; the error is small at the gross macroscopic level and increases with magnification so that it becomes very great in electron microscopy.

In diagnostic practice the problem arises both from the heterogeneous composition of organs, tissues and cells, and from the tendency of many diseases to give rise to focal rather than diffuse changes. To take an extreme example, a needle biopsy of kidney that contains no glomeruli is a manifestly inadequate sample for the investigation of glomerular disease. Similarly, a renal biopsy containing only a single normal glomerulus does not entirely exclude the presence of a focal glomerulonephritis.

Sampling error at the macroscopic level is most simply illustrated by the detection of hepatic metastases at autopsy. In an extensively infiltrated liver, tumour may be visible on the cut surface of every slice; deposits may even be visible at the capsular surface. In another case one may wish to guarantee the detection of a solitary hepatic metastasis, 1 cm in diameter. Without resorting to statistical analysis it would be fair to say that it would be improbable that a single cut through an adult liver would

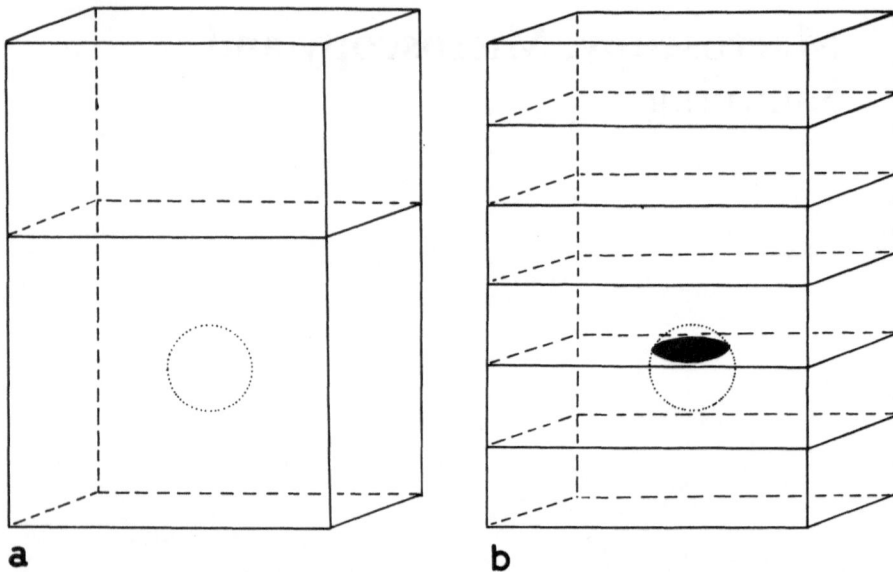

Fig. 2.1a, b. Schematic illustration of sampling error in the detection of a solitary lesion (*sphere with dotted outline*) in an organ (*rectangular form*) by slicing. **a** If D is the diameter of the lesion and L is the longitudinal dimension of the organ, the probability of transecting the lesion with a single transverse slice is given by D/L. **b** If multiple slices are made, each of the thickness $<D$, the lesion is certain to be detected on one or more cut surfaces.

transect a solitary metastasis 1 cm in diameter. However, if the liver was cut into slices less than 1 cm in thickness the metastasis is certain to be transected (Fig. 2.1).

A consistent sampling method must be strictly adhered to when attempts are made to determine the incidence of occult neoplastic or preneoplastic disease accurately, as has been done for occult primary carcinomas of the thyroid (Silverberg and Vidone 1966) and prostate (Franks 1954). In routine diagnostic practice it is often impractical to sample a specimen as extensively as one would when following a research protocol. An acceptable compromise as to the minimum number of blocks and sections must be decided upon. If one is limited to a single block (liver, renal, bone trephine biopsies, etc.), then it may be important to examine sections taken at intervals through the block; intermediate sections should be kept for special stains as required.

A factor which limits the accuracy of biopsy techniques is the size of each sample in relation to the object as a whole. To take the liver as an example again, a section from a generous needle biopsy represents approximately one five-millionth of the liver; the potential sampling error in needle biopsies of the liver is therefore very great. The problem is even greater at the ultrastructural level. For electron microscopy a typical thin section would be only one of about 2×10^9 that could be prepared from an entire human liver; and allowance must also be made for the grid bars that obscure parts of the sections when they are actually examined in the electron microscope.

The flea scurries about and his whole territory is less than one square yard of a dog. The electron microscopist can be very busy too, and in his whole lifetime not cover as much territory (**Pease 1964**).

Sampling Biopsies and Surgical Resections

Any biopsy or resection specimen should be systematically dealt with in such a way as to satisfy three needs.

1. To provide a record of the gross appearance of the object, the location of lesions, their dimensions, contour, colour, shape and texture. Due allowance may have to be made for discoloration due to fixation and other artifactual distortions if the tissue is not received fresh.
2. To sample the object adequately to obtain meaningful and representative blocks for histology.
3. To provide for special investigations such as immunohistology, electron microscopy and tissue culture.

If the tissue is received unfixed, as it ideally should be, the pathologist must realise the opportunity to preserve at least some fresh tissue, when indicated by the circumstances of the case, for special investigations that cannot be undertaken on routinely fixed tissue. For example, if infection is suspected then a piece of fresh tissue should be sent for microbiological investigations; if the specimen is a lymph node and lymphoma is suspected, fresh tissue should be snap-frozen for immunohistology; if the specimen is a renal biopsy or possibly a tumour in which diagnostic difficulties can be anticipated, fresh tissue should be fixed for electron microscopy (e.g. in glutaraldehyde rather than·formalin); fresh breast tumours should be sampled for oestrogen receptor assays. Also, if a permanent record of the gross appearance is needed, photography should be done while the tissue is still fresh. It is therefore useful to run through a simple check-list of options for realising the full diagnostic and investigative potential of fresh tissue whenever a specimen is received:

— Photography while the specimen is still intact
— Imprints for cytology
— Tissue for microbiology where infection is suspected
— Tissue for immunohistology, especially cell-surface markers which cannot easily be detected in paraffin sections of fixed tissue
— Tissue for electron microscopy
— Tissue for histochemistry
— Tissue for biochemistry (e.g. steroid receptor assays)

Having sampled the fresh tissue for any special investigations, the specimen can then be fixed. Ten per cent formalin (i.e. 4% formaldehyde), buffered or unbuffered, is routinely used in most laboratories, but for some lesions other fixatives may be more appropriate. For example, Müller's chromate fixative is preferred for suspected or known phaeochromocytomas since it will give a positive chromaffin reaction contributing to the diagnosis.

Calcified tissues and lesions may be processed either in their calcified state, requiring embedding in plastic or resin (mandatory for bone trephines for the investigation of metabolic bone disease), or after decalcification.

If a lipid storage disorder or adipose tumour is suspected some of the fixed tissue should be kept back for fat stains. Processing into paraffin wax will remove fat from the tissue when it is exposed to alcohol.

The actual procedure adopted will be dictated by the nature of the individual case and local anatomy. The provisional diagnosis determines the general way in which the tissue is handled; the specimen may well have to be re-examined and re-sampled if the provisional diagnosis is changed by the initial histological findings.

The "cut up" has the reputation of being the most mundane and lowly task of the pathologist. Though it lacks the glamour of the surgical operation or even the endoscopy clinic, the patient is best served by the pathologist who integrates gross and microscopic information to form a diagnosis than by one who rarely steps outside the two-dimensional world of stained tissue sections.

Some general indication as to the best way to handle tissue is given below; Kennedy (1977), Pierson (1980) and Rosai (1981) give comprehensive guides. What actually happens will very much depend upon the custom and practice in individual laboratories, but the principles are, I feel, best conveyed in the didactic style of what follows.

Suspected or Proven Neoplasia

Surgical resections for neoplasia must be accurately described and measured, preferably having received the specimen fresh. It is useful to draw or photograph large specimens with sites of the blocks taken marked out. Histology blocks should be taken to answer four questions (Fig. 2.2).

1. What is the nature of the tumour?
2. What is the extent of spread?
3. Is local resection adequate?
4. Are there any associated, possibly precancerous, lesions?

1. *Tumour.* Blocks that include the edge of a tumour are invariably more informative than less well-vascularised central samples showing sclerosis or necrosis. Vascular and lymphatic permeation is often more evident peripherally and the lymphoreticular reaction, which may have prognostic significance, is often strongest at the edge. Some tumours are, or appear to be, encapsulated. Histological confirmation of a capsule and its integrity is useful, for example, in the interpretation of thyroid nodules. Blocks from the edge of gut tumours may show a local origin from adjacent mucosa.

Sometimes a single block of tumour will suffice, but some tumours are so heterogeneous that a small sample may convey a false impression of uniformity. For example, testicular tumours that superficially appear to be purely seminomatous should be extensively sampled to exclude the presence of teratomatous or trophoblastic elements.

Problems sometimes arise when the tumour in a mastectomy specimen is difficult to locate, or when it is desirable to assess its distribution precisely. Breast specimens are easier to examine if the fat is hardened by cooling so that thin slices can be cut (Davies et al. 1973). In some cases this approach has given valuable information about the possible multicentricity of tumours (Gallagher and Martin 1969).

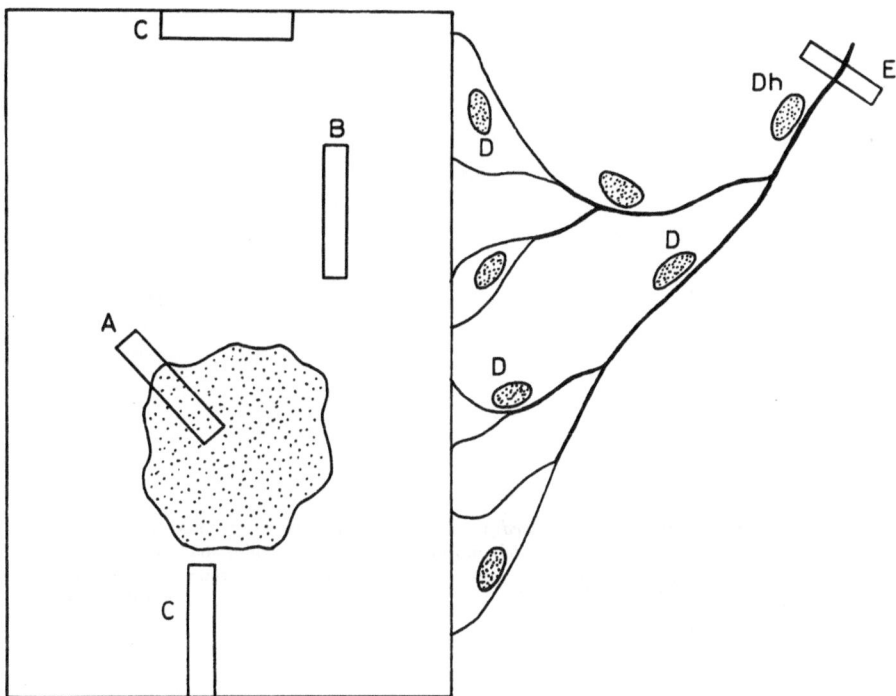

Fig. 2.2. Schematic protocol for adequate sampling of a resection specimen for proven or suspected neoplasia. *A*, tumour including full depth and lateral edge; *B*, mucosa or parenchyma remote from the tumour; *C*, resection margins; *D*, representative sample of regional lymph nodes; *Dh*, highest lymph node; *E*, vascular pedicle.

The results of painstaking examination of some specimens can have immediate therapeutic implications. This is particularly so with the assessment of splenectomy specimens from staging laparotomies in patients with Hodgkin's disease. Unless involvement is grossly obvious, the entire organ should be carefully cut into 2–3 mm slices and obvious deposits or large Malpighian bodies selected for histology (Farrer-Brown et al. 1972).

2. *Vascular Pedicle.* Vascular permeation can often be seen in sections of tumours, but just as important is the recognition of gross permeation at a macroscopic level. Some tumours, such as hypernephromas, have a propensity to spread along veins; this is best sought for by taking blocks of the entire vascular pedicle. Likewise, several blocks of spermatic cord should be examined from every orchidectomy performed for neoplasia.

3. *Lymph Nodes.* Most routine methods of lymph node sampling are barely adequate. Monroe (1948), in a meticulous study, cleared the axillary fat from mastectomy specimens in oil of wintergreen after dehydration in alcohol. As a result he was able to recover 2643 lymph nodes in total from 87 radical mastectomies, an average of 30.3 nodes per specimen. About 50 lymph nodes can be recovered from standard colorectal

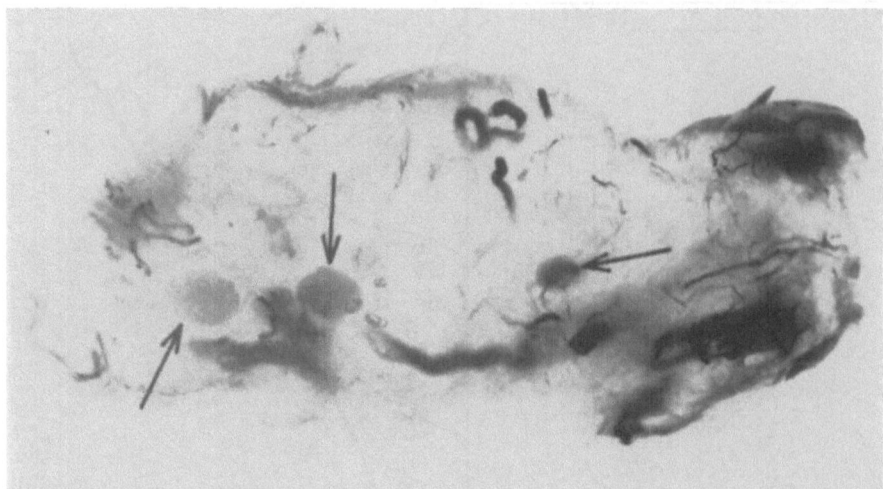

Fig. 2.3. Identification of lymph nodes (*arrowed*) in slices of adipose tissue (e.g. mesentery, axillary fat) can be facilitated by clearing the tissue after dehydration in alcohol. Transillumination allows even the smallest nodes to be isolated. This theoretically attractive method is inordinately time-consuming for most laboratories; any routinely adopted method should yield a fully representative sample of the nodes that could be demonstrated by this technique. $\times \frac{1}{2}$

resections if the mesentery is dehydrated and cleared (Berlin and Brines 1951) (Fig. 2.3). An alternative method, adopted by Berlin and Brines, is to embed the entire specimen in Carbowax, a water-soluble synthetic wax, and then cut thin slices from which the nodes can be retrieved for histology. Although these methods are probably too cumbersome for routine use, it is important to avoid the opposite extreme. Lymph node involvement is often detected in a somewhat arbitrary way; the recorded frequency of nodal metastases may rely heavily on the inquisitiveness and patience of the individual pathologist. It is essential that a standard sampling method is followed. Selection of nodes by palpation is to be avoided because it biases the sampling in favour of involved nodes. A better method is to slice the mesenteric or axillary fat at regular intervals and process all nodes presenting at the cut surfaces.

Having decided on the method to be used for detecting lymph nodes, there remains the question of sectioning. The routine practice in many centres is to examine a single random or equatorial section of each node. Saphir and Amromin (1948) examined 149 axillary lymph nodes recovered from 30 mastectomies. Serial sections were taken through each lymph node block, an average of 332 sections per node. Nodal metastases were found in ten out of the 30 mastectomy specimens, all of which had previously been reported to be free from metastases on routine assessment. Lymph node size and consistency proved to be unreliable indicators of involvement.

The problem of detecting a small deposit of tumour in a single node has been examined mathematically by Wilkinson and Hause (1974). They estimate a 30% error for most routine methods. The detection rate depends on several variables such as the size of the lymph node, the size and location of the deposit of tumour, and the number of sections examined. Wilkinson and Hause calculate that, in seeking a 1 mm diameter peripheral metastasis in a 10 mm diameter node, only 13.9% success could be expected on a single equatorial section. The success rate increases to 40.4% if two quarter and centre sections are studied. A 1 mm diameter peripheral deposit in a 4 mm diameter node can be picked up with a 39.5% success rate on a single equatorial section, increasing to 100% success with two quarter and centre sections (Fig. 2.4).

Overt nodal involvement by tumours obviously conveys a worse prognosis than does the total absence of metastases, but the precise clinical significance of *occult* nodal metastases (i.e. discovered only by serially sectioning lymph nodes) remains doubtful. The 5-year survival prospect for breast carcinoma patients found to have occult nodal metastases is almost as good as that for patients whose nodes are known to be entirely free from tumour (Pickren 1961). Clearly, therefore, the information obtained from an exhaustive search for very tiny tumour deposits is going to be of proportionately marginal clinical value. More important is the detection of tumour in a solitary node removed from a patient with known or suspected neoplastic disease. It is in these circumstances that the data of Wilkinson and Hause provide an insight into the limitations of perfunctory methods of sampling and histological assessment.

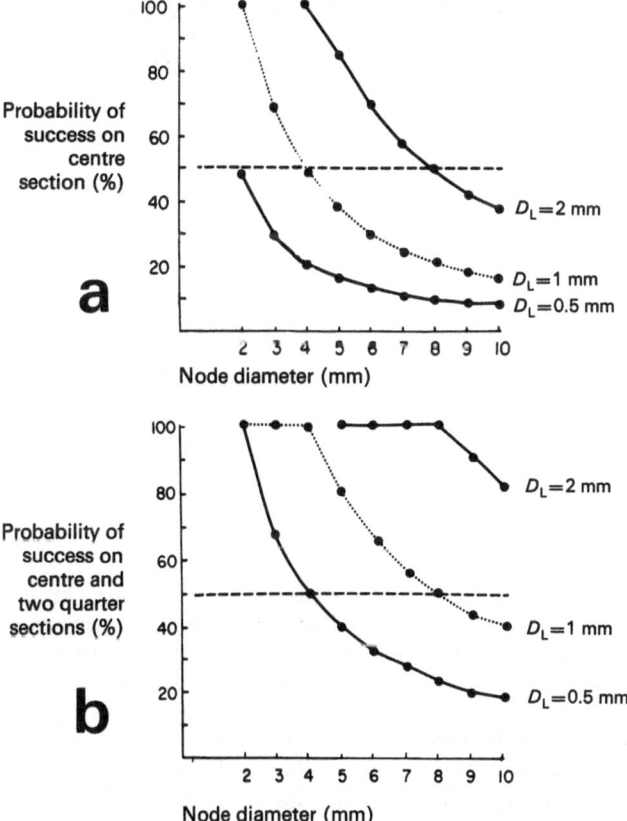

Fig. 2.4a, b. Mathematical treatment of the histological search for a solitary tumour deposit in a single lymph node. a The probability of finding a deposit with a single equatorial section is compared with b the substantially increased probability of success if the node is divided into four slices of equal thickness before sectioning. D_L, diameter of a randomly distributed solitary tumour deposit, (Wilkinson and Hause 1974)

4. *Resection Margins*. The assessment of resection margins carries a considerable degree of potential sampling error; a single section can easily miss a slender tract of tumour cells or a focus of lymphatic permeation. With gut resections, blocks are usually taken randomly from proximal and distal margins. Blocks from the bronchial resection margin should be taken from pneumonectomy and lobectomy specimens containing tumours. Skin ellipses bearing tumours should be sectioned across the transverse axis and longitudinal blocks prepared from the remaining material. The end of the ureter from ureteronephrectomy specimens removed for transitional-cell tumours of the renal pelvis should be examined to exclude in situ urothelial neoplasia spreading distally. Polypoid tumours must be sectioned through their base or stalk.

5. *Detection of Associated Lesions*. Blocks taken from the edge of a tumour are more likely to lead to the detection of local precancerous or other pathogenetically associated lesions. A diffuse process, such as ulcerative colitis, should be macroscopically evident; blocks taken randomly are then sufficient to confirm the diagnosis. Focal proliferative lesions in breast biopsies and mastectomies warrant extensive sampling. Subgross methods can be used to pick up focal abnormalities in thin slices that have been cleared prior to stereomicroscopic examination. A "Swiss roll" (coiled mucosal strip) technique is an economical method for the examination of gastric mucosa from gastrectomy specimens (Kennedy 1977); this can be used to detect dysplasia, gastritis, and metaplasia in stomachs resected for cancer.

Ischaemic and Vascular Lesions

Biopsies or resections for vascular lesions need to be assessed in such a way as to answer three questions.

1. Where are the vascular lesions located?
2. What is their exact nature?
3. What are the consequences?

Obviously, the vascular lesion must be located grossly before its nature can be determined histologically. It is equally obvious that only a very short segment of a vessel needs to be occluded, even partially, for widespread ischaemic damage to occur distally. If, for example, a length of ischaemic small intestine is removed, a single random tissue block containing mesenteric vessels would be a hopelessly inadequate way of detecting obstructive vascular lesions. Ideally the vascular bundle should be cut transversely at close intervals and any grossly visible occlusions selected for histology. In the absence of macroscopically detectable lesions, a series of random blocks should be taken.

The ischaemic effects of vascular occlusion are usually macroscopically obvious, but it is useful to have histological confirmation and possibly establish the age of the lesion. More important is the viability of the resection margins, within the limitations of morphological criteria.

Infection and Inflammatory Lesions

Surgical resections for specific infective processes are not common because antibiotic therapy is usually employed as the first treatment of choice. Sometimes obvious infective lesions are discovered unexpectedly during surgical procedures. On other occasions tissue is removed from patients in whom infection is not the most likely diagnosis, but should certainly be considered as a remote possibility. It is vital in these cases that a portion of the fresh lesional tissue is taken for microbiological studies before fixation. This is important, for example, in cases where tuberculosis is suspected, because cultures give higher detection rates than Ziehl-Neelsen staining of tissue sections.

In some infections the causative organisms can be so sparse as to elude detection by microscopy. This is certainly often the case with tuberculosis. In actinomycosis the quantity of exuberant granulation tissue tends to be disproportionately excessive. If the diagnosis is suspected and organisms are not seen on initial sampling, more of the lesion should be processed for histology before some other explanation for the appearance is sought.

Resections for chronic inflammatory bowel disease need careful sampling by taking multiple blocks and recording their anatomical location. For example, in Crohn's disease the lesions are patchy and some of the morphological hallmarks, such as granulomas, are often very sparse. Furthermore, resection margins need careful histological assessment.

Squashes of fresh biopsy tissue are a useful way of looking for schistosome ova (Rosenberg and Black 1959); staining is unnecessary (Fig. 2.5).

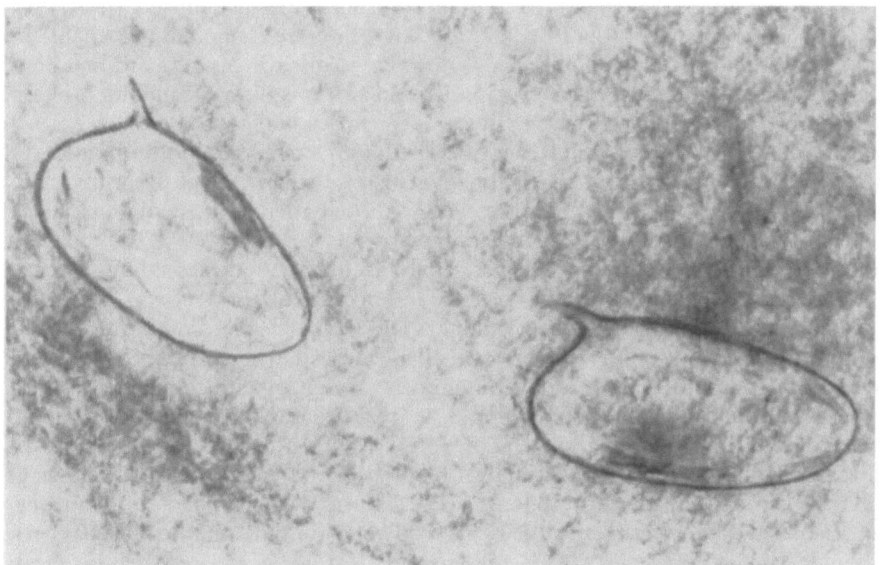

Fig. 2.5. Unstained squash preparation of a fresh rectal "snip" biopsy showing ova of *Schistosoma mansoni*, each endowed with the typical lateral spine. Fresh liver biopsies can be similarly treated. × 250

Metabolic and Degenerative Diseases

Although these constitute a relatively small proportion of the histopathologist's practice, the tissue must be handled with special regard to the nature of the diagnostic problem. Sampling itself is not difficult because most metabolic and degenerative disorders tend to produce relatively diffuse morphological changes. The fundamental rule is to process the tissue so as to retain any abnormal substances, such as urates for which the tissue should be fixed in alcohol because urates are soluble in aqueous fixatives, calcium pyrophosphate which is eliminated by decalcification, fat which is removed by alcohol and other fat solvents, and excess glycogen in liver which is water soluble.

Where histochemical methods are required, with muscle biopsies for example, fixation in formalin is generally to be avoided; frozen sections cut from fresh tissue are invariably used.

Specific Types of Biopsy

Needle Biopsies

Since their introduction in the 1930s (Baron 1939), percutaneous needle biopsies of liver and kidney have revolutionised our understanding of hepatic and renal diseases. Needle biopsies have also been successfully used for the diagnosis of lymph node enlargment (Meatheringham and Ackerman 1947), thyroid disorders (Hawk et al. 1966) and breast lumps (Saltzstein 1960). In most instances needle biopsies are a valuable investigative method provided that sufficient material is obtained to enable a diagnosis to be made with confidence. False-negative results will occur if a focal lesion is missed by the needle. Diagnosis by rapid frozen sections of excisional biopsies is the next resort in breast lesions where clinical suspicion of malignancy remains.

Despite misgivings voiced about the risk of tumour dissemination by needle biopsies, there is no objective evidence that patients whose tumours have been subjected to pre-operative needle biopsy fare less well than those patients whose tumours have not been biopsied in this way (Berg and Robbins 1972).

For obvious reasons sampling error in renal and hepatic biopsies is low for the detection of diffuse changes but high for focal lesions. Kellow et al. (1959) examined 103 needle tissue samples from autopsy kidneys and reported a detection rate of 84% for diffuse changes, but only 51% for focal changes. Similarly, correlation between liver biopsies and subsequent autopsy findings is strong for diffuse hepatic lesions and weaker for focal changes (Wagoner et al. 1951). Sampling error was claimed to be largely responsible for a correlation rate as low as 33% in the diagnosis of cirrhosis in simultaneous or sequential liver biopsies from the same patient (Soloway et al. 1971). In a detailed study, Baunsgaard et al. (1979) examined simultaneous percutaneous liver biopsies from 70 patients. All were from the right lobe, separated by about 5 cm. Cholestasis, steatosis and cirrhotic features correlated well between biopsies, but focal changes such as ductular proliferation and acidophilic bodies were poorly duplicated. Sampling error leading to clinically significant consequences can be minimised by taking multiple liver biopsy cores with no increased risk to the patient (Maharaj 1986).

The risks of missing the lesion or target organ can be substantially reduced if the insertion of the needle is done under radiological control. This is unnecessary for the liver and kidney, but common practice with focal lung lesions.

A renal biopsy is almost useless unless it contains cortical tissue; it is claimed that this can be rapidly ascertained by doing imprints of the tissue core (Meyer 1973), though examination by dissecting microscopy may be just as effective in experienced hands.

Mucosal Biopsies

Mucosal biopsies are now commonplace in the investigation of patients with gastrointestinal disease; modern flexible endoscopes have considerably extended the anatomical range of the technique.

Dermatologists have to be the gross pathologists of the skin; endoscopists are now assuming a similar role in internal organs; there may be some benefit in pathologists being observers in clinics that conduct these examinations regularly. Some lesions may be difficult to delineate in the fibreoptic image, but advances in endoscopic technique, notably in Japan (Kawai et al. 1979), include spraying the mucosal surface with dye, either to enhance the relief of lesions on the mucosa or to stain them. Advocates claim that this adds a functional dimension to endoscopy and may assist in selecting very small lesions for histology.

Knowledge of the exact site of an endoscopic biopsy is important. In the oesophagus this is usually given as the distance from the incisor teeth; glandular mucosa in the distal oesophagus may have different significance at different levels. A gastric biopsy that has an antral mucosal pattern would, of course, be normal in the antrum, but might reflect metaplasia if the biopsy had been taken from the body.

If orientation is important, and one is doubtful that it has been done correctly, a single section can be cut from the block with minimum trimming to check it. The wax block can be melted down and the tissue rotated before re-embedding should this prove necessary. Tissue should not be wasted by cutting deeper into the block when reorientation of the biopsy is all that is required.

Capsule biopsies of jejunal mucosa for the investigation of malabsorption states can be assessed by stereomicroscopy. The architecture of the villous surface is clearly visualised by this method, but this procedure is not so much in vogue as formerly. However, any disparity between the stereomicroscopy and the villous architecture seen in a section may be resolved by examining further sections taken at deeper levels into the tissue block. Proper orientation of any mucosal disc always makes interpretation easier (Fig. 2.6).

A biopsy of adequate depth is essential for the diagnosis of certain disorders. Rectal biopsies done for the diagnosis of amyloid and vasculitic problems (e.g. polyarteritis nodosa) should ideally include a substantial amount of submucosa. Superficial rectal biopsies devoid of submucosa are inadequate for the diagnosis of Hirschsprung's disease. Colorectal Crohn's disease can be difficult to diagnose with certainty on superficial biopsies because some of the diagnostically useful stigmata are often found only in the submucosa and deeper layers (Fig. 2.7).

The success rate for endoscopically diagnosing malignancy in ulcerated mucosal lesions depends on the number of biopsy samples taken and their site. Hatfield et

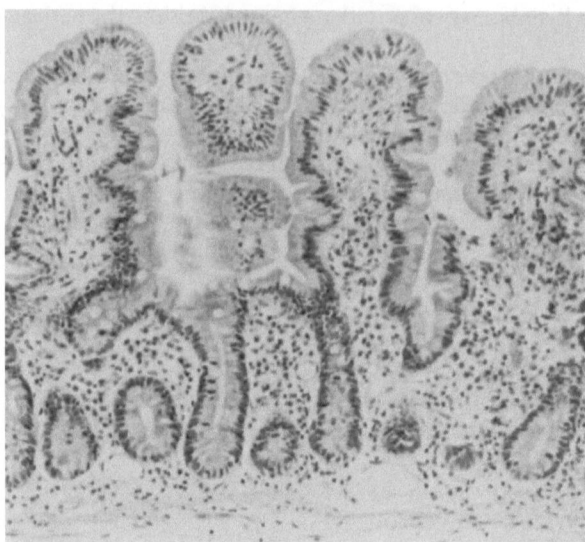

Fig. 2.6. Accurate assessment of skin and mucosal biopsies is easier if the specimen is correctly orientated in the wax block. In this jejunal biopsy the plane of section passes vertically down through the villi and crypts, the optimal plane for assessment of villous architecture. Haematoxylin and eosin. × 140

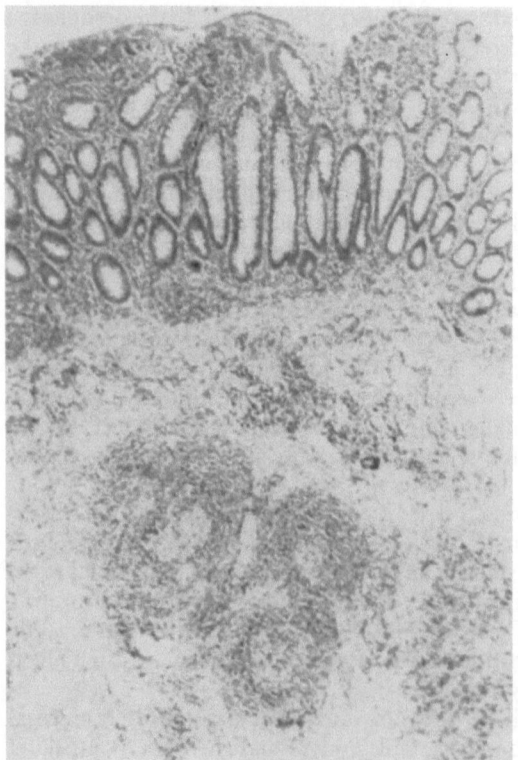

Fig. 2.7. Rectal biopsy from a patient with chronic inflammatory bowel disease. Foci of inflammation in the sub-mucosa favour Crohn's disease. Interpretation of the appearances would have been more difficult if the biopsy had consisted only of mucosa. Haematoxylin and eosin. × 28

al. (1975) sampled gastrectomies for proven ulcerated carcinomas and showed that it was necessary to biopsy both the slough and ulcer rim for the maximum pick-up rate. Whether multiple or single fragments are submitted for examination, it is important to examine several sections through the block. Multiple fragments may not all be embedded at the same level in the block; small ones may be missed altogether in single sections.

Curettings

Curetted specimens are awkward to orientate, but this is often only of importance with skin lesions. While there is little problem in identifying simple warts in curetted samples, the distinction between inverted follicular keratosis, keratoacanthoma, squamous-cell carcinoma and other lesions may be difficult; the distinction between keratoacanthoma and squamous-cell carcinoma in such biopsies may be impossible. Multiple sections may sometimes enable the architecture of the lesion to be pieced together.

Uterine curettings are usually obtained in a sufficiently small quantity to justify embedding at least most or even all of the tissue submitted. Endoscopic resections of bladder and prostatic lesions, on the other hand, commonly produce a vast quantity of tissue. Because one is limited to examining only a proportion of the material, it is preferable to select particular fragments macroscopically than to process a random sample. With bladder resections, the more solid portions will provide therapeutically useful information about the extent of tumour invasion. Hard whitish chips from transurethral resections of prostate are more suggestive of malignancy than the rubbery fragments which often merely show benign hyperplasia.

Cone Biopsies

Sampling protocols for cone biopsies of the cervix are well established in most laboratories, though Rubio et al. (1978) have suggested that there is still room for further improvement in international standardisation. For epidemiological purposes, reports should include information about cone size and the number of sections examined. Nichols et al. (1968) advocate step-sterial sections; in their investigation of 624 cone biopsies that had been routinely reported, these extra sections gave additional information in 105 cases.

Specimen Radiology

No doubt owing to the popularity of mammography, radiographic examination of resections and biopsies has been almost entirely confined to breast disease. Microcalcification sometimes marks the site of a neoplastic lesion in the breast; radiology can be used to locate exceedingly small lesions that may not be apparent on palpation

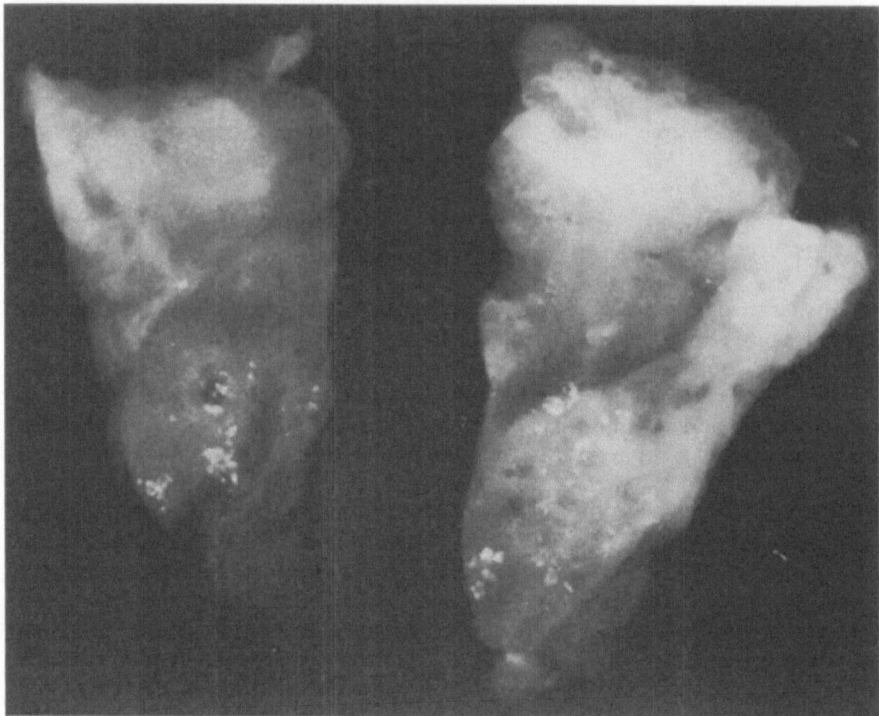

Fig. 2.8. Radiograph of breast tissue slices containing an invasive ductal adenocarcinoma. The flecks of microcalcification were subsequently shown to be present in the necrotic intraductal component and in periductal stroma. (From a specimen radiograph supplied by Dr. J. D. Davies) × 2.5

or slicing the specimen (Fig. 2.8). In a large study (Gallagher and Martin 1969), 2371 breast tissue samples were examined and, after examination by routine pathological methods, 1659 were deemed to be free from carcinoma. Radiographic examination of these tentatively benign biopsies revealed focal microcalcification in 375; this had not been noted on previous routine histology in 183 instances. Additional tissue from these cases disclosed 11 ductal or lobular carcinomas that were otherwise occult. Routine breast biopsy radiography would be a formidable burden for many laboratories, but should certainly be done when initial sampling fails to reveal a lesion that has been picked up on mammography.

Specimen radiography can also be helpful when a portion of bone, excised for a radiologically evident lesion, shows no expansion or external distortion that would enable the lesion to be macroscopically located without mutilating the specimen.

Section Thickness

Standard sections are routinely cut at 5 μm, sufficiently thin to impart transparency to the tissue and minimise cellular superimposition. The tissue is thereby rendered essentially two-dimensional. The 5 μm standard section is a compromise that satisfies most diagnostic purposes. Although there may be some reluctance to deviate from the familiar standard section, sections thicker or thinner than 5 μm can be very useful.

Thick Sections

The purpose of cutting thick (>5 μm) sections is to reduce sampling error. A 20 μm section has four times the volume of a standard section. Although virtually useless for diagnosis of most lesions because superimposition obscures detail, it can be a useful way to find pathogens or fine particles that are sparsely distributed. Thus, thick sections can be used to search for acid-fast bacilli in suspected tuberculous lesions or asbestos bodies in fixed lung tissue (Fig. 2.9). The microscope must be focused through all planes of the section during the examination.

Thin Sections

Thin sections (<5 μm) can be cut from ordinary paraffin wax blocks but it is technically much easier to use tissue embedded in harder media such as digol distearate, epoxy resins or methacrylate (Te Velde et al. 1977). These media provide greater support for the relatively flimsy sections. Current popular applications include renal biopsies, lymph node biopsies, and bone marrow trephines (Fig. 2.10). The superb clarity afforded by 1 μm plastic sections is due to the virtual elimination of cellular superimposition and to the very considerable reduction in tissue shrinkage as compared with paraffin processing.

Plastic embedding has a further advantage for bone marrow trephines in that less decalcification is required. Stains may have to be adapted to work successfully on plastic sections; the range of possible stains is wide and even immunohistology can be performed successfully.

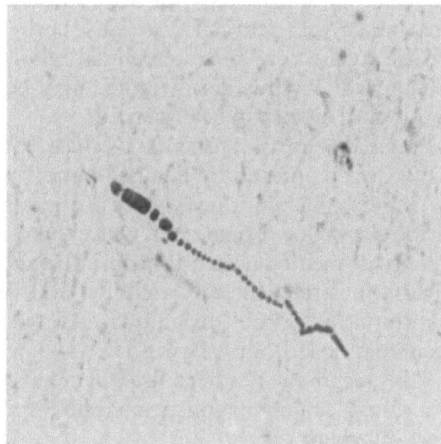

Fig. 2.9. Ferruginous (asbestos) body in an unstained 20 μm thick section of a lung tumour. The beading is characteristic. $\times 558$

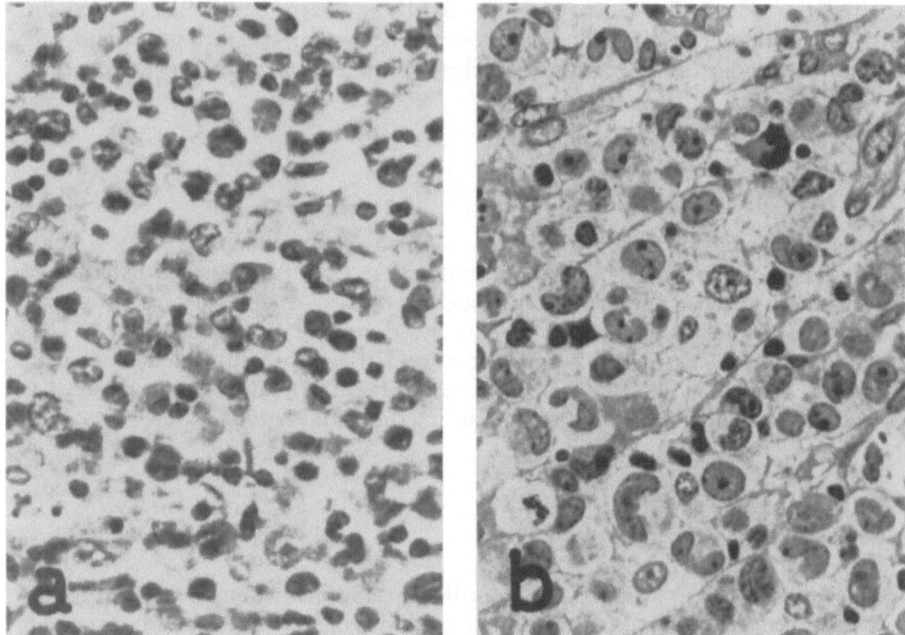

Fig. 2.10. Nodal involvement by mycosis fungoides. One portion (a) of the formalin-fixed lymph node was processed for paraffin sections (approximately 5 μm); another portion (b) was processed for thin plastic sections (approximately 1 μm). The latter has considerably greater cytological clarity. Haematoxylin and eosin, × 558

Imprints and Smears

While tissue sections have long been the predominant diagnostic medium of the histopathologist, these can be usefully supplemented by cytological methods. Tissue imprints and smears have the advantages of speed because tissue processing is eliminated, and cytological clarity because entire cells are spread out rather than transected. Tissue architecture and intercellular relationships are disrupted by imprint and smear techniques, but this disadvantage may be compensated for by the ease with which cytological detail can be discerned.

Rapid diagnosis of tumours by the preparation and examination of cells from fresh tissue was pioneered by Dudgeon and Patrick (1927). Tissue imprints are prepared by simply touching the exposed external and/or cut surfaces or fresh tissue with a clean glass slide. Some cells will adhere to the slide and they can be stained by a variety of methods after appropriate fixation. The technique and typical results are vividly illustrated in an article by Godwin (1976). A combination of scraping and imprinting ("scrimp" technique) may yield a larger harvest of cells for cytological examination (Abrahams 1978).

This technique has been used by many pathologists as an adjunct to histology and is a valuable rapid diagnostic method for a wide range of tissues and lesions (Blaustein and Silverberg 1977).

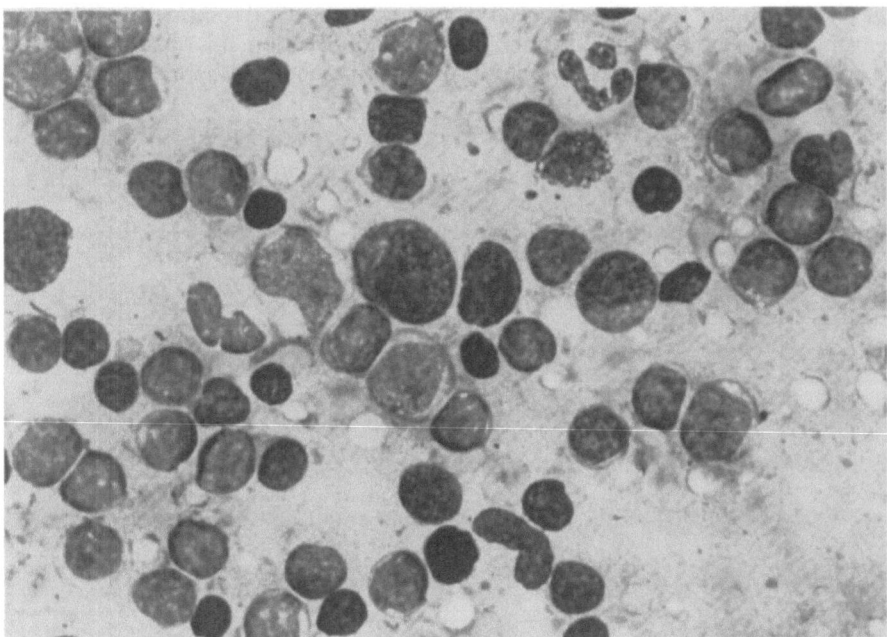

Fig. 2.11. Imprint prepared from the fresh cut surface of a lymph node from a patient with infectious mononucleosis. The cytological appearances can be easily compared with blood and marrow smears from the same patient. Giemsa, × 880

Tissue imprints can be useful for intraoperative diagnosis and can sometimes clarify an equivocal frozen section appearance. It seems to be particularly useful where tissue distortion by chronic inflammation may simulate the architecture of a malignant neoplasm in frozen sections (e.g. chronic pancreatitis versus adenocarcinoma), or where a small focus of malignancy may be overlooked because most of the area of a frozen section is occupied by necrosis or inflammation (imprints can sample a greater area of tissue surface than frozen sections can in a limited time), or where the submitted biopsy is too small to enable frozen sections to be cut (Suen et al. 1978).

Imprints are also a useful aid to lymph node biopsy interpretation and lymphoma classification (Fig. 2.11). Well-preserved cytology can be easily compared with blood and marrow smears from the same patient (Moore and Reagan 1953).

Information and Magnification

The examination of tissues by histology is the chief diagnostic method of the histopathologist, but interpretation does not necessarily become either easier or more accurate with increasing magnification. A common misconception among trainees is that the diagnosis that seems elusive at the light microscope will always be easier at the electron microscope. Examination of a biopsy or surgical specimen at different magnifications gives, in fact, quite different sorts of information. It is possible to draw up a spectrum of the type of information that can be obtained by macroscopic or microscopic means (Table 2.1).

Table 2.1. Examples of information obtained from examination of tissues at different magnification ranges

Relative magnification	Method	Typical information
1	Unaided eye	Size, shape, texture colour, and contour of gross lesions
<40	Dissecting microscope	Mucosal architecture in jejunal biopsies, etc.
25–250	Light microscope	Tissue architecture, vascular pattern, differentiation of tumours, etc.
250–1000	Light microscope	Mitotic activity, pleomorphism, etc.
		Pathogens—fungi, protozoa, bacteria, viral inclusions
>1000	Electron microscope	Organelles, desmosomes, microvilli, etc. Viruses

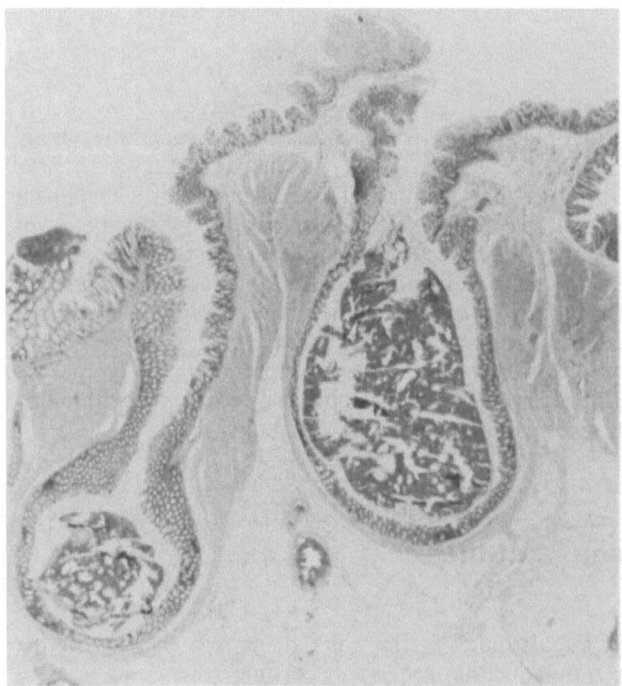

Fig. 2.12. Diverticular disease of the colon. The characteristic mucosal herniations through the muscularis are readily seen at this very low magnification. Haematoxylin and eosin, ×6

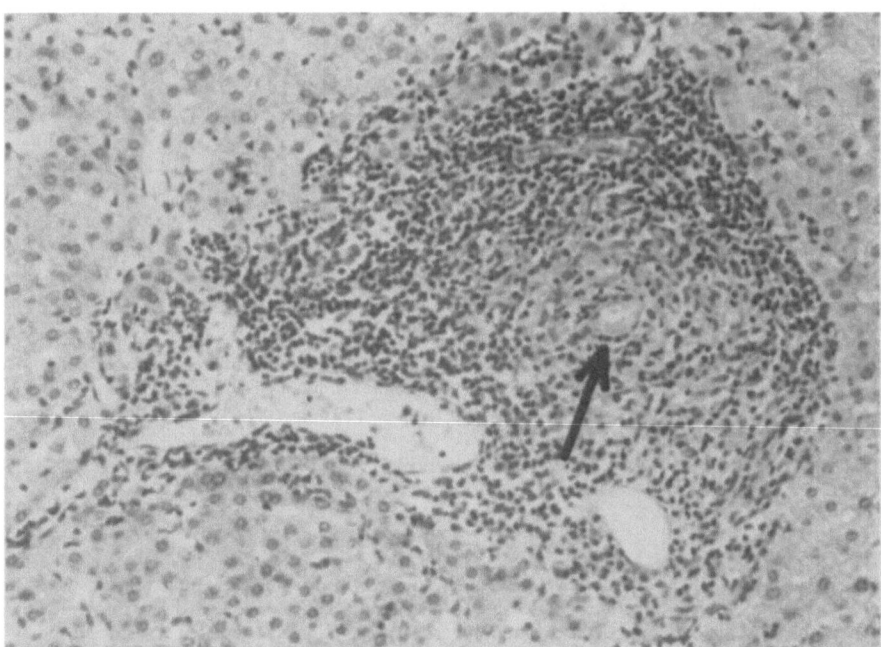

Fig. 2.13. Primary biliary cirrhosis. A granulomatous aggregate of epithelioid macrophages forms a mantle around a degenerate interlobular duct. Lymphocytes densely infiltrate the portal fibrous tissue. Medium- or high-power light microscopy is necessary for the accurate recognition of these changes. Needle biopsy. Haematoxylin and eosin, × 215

The relevance that this concept has to specific diagnostic problems is illustrated by the following examples.

Example 1. A short length of sigmoid colon removed from a patient with rectal bleeding and intermittent lower abdomen pain showed numerous typical diverticula; the diagnosis of diverticular disease of the large bowel was not in doubt. Low-power light microscopy of sections, usually prepared routinely, will confirm the presence of mucosal herniation through the muscularis (Fig. 2.12). Melanosis coli, recognisable only at higher magnification, would certainly be an additional finding of interest, but would not necessarily facilitate the actual diagnosis of diverticular disease. Electron microscopy would contribute nothing to the diagnosis. This naively simple example illustrates the importance of gross examination and low-power microscopic interpretation in the diagnosis of many lesions. Microscopy was not, in fact, particularly contributory in this case.

Example 2. A needle liver biopsy was done on a patient with pruritis, jaundice and an antimitochondrial antibody. The unaided eye contributed very little, other than to record the size and colour of the biopsy. Low-power microscopy revealed moderate architectural distortion. However, medium-power microscopy clearly showed duct destruction, granulomas and expansion of portal tracts (Fig. 2.13), appearances which in this clinical context can be interpreted as primary biliary cirrhosis. As in the first example, the electron microscope has little to offer since there are no specific ultrastructural features that would resolve the diagnosis if it were in doubt. In this example, the maximum amount of useful information was obtained by medium-power light microscopy.

Example 3. A right upper lobectomy was done on a patient with weight loss, haemoptysis, and a highly suspicious pulmonary lesion visible radiographically. The lobectomy specimen contained a greyish-white lesion 3.5 cm in diameter, apparently neoplastic and probably malignant. Invasion, pleomorphism and mitotic activity were visible microscopically in this neoplasm, but the exact histogenesis of such a tumour could be established only by the identification of some structural or biochemical marker of differentiation.

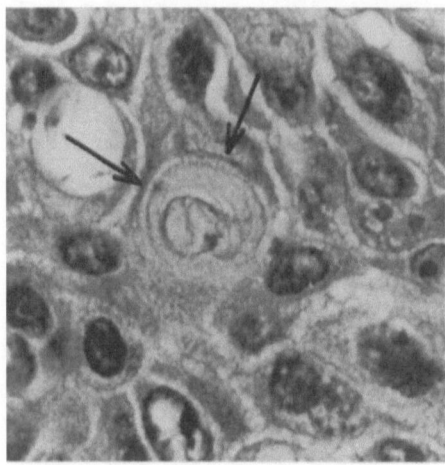

Fig. 2.14. Poorly differentiated squamous-cell carcinoma of lung. Oil-immersion light microscopy reveals numerous intercellular bridges ("prickles") consistent with a squamous histogenesis. Haematoxylin and eosin, × 880

In this case, high-power light microscopy reveals intercellular bridges sufficiently numerous to justify a diagnosis of squamous-cell carcinoma (Fig. 2.14). Ultrastructural confirmation of desmosomes in this tumour would be useful, but not perhaps essential. Relevant information was obtained across the magnification range in this example. Gross appearances were interpreted as those of a tumour, low- and medium-power light microscopy showed features of a malignant neoplasm, and high-power light microscopy established its histogenesis.

Example 4. A young woman had a left lower lobectomy to remove a circumscribed tumour approximately 5.0 cm in diameter; this had given rise to haemoptysis. Light microscopy showed a cellular tumour devoid of obvious glandular or squamous differentiation, and it did not have the pattern of an oat-cell carcinoma. The possibility that it was a carcinoid tumour was considered; the cells proved to have argyrophylic cytoplasm on light microscopy, and electron microscopy showed numerous electron-dense membrane-bound granules morphologically consistent with a tumour of APUD lineage (Fig. 2.15). Electron microscopy was necessary to obtain the most crucial item of information about this lesion.

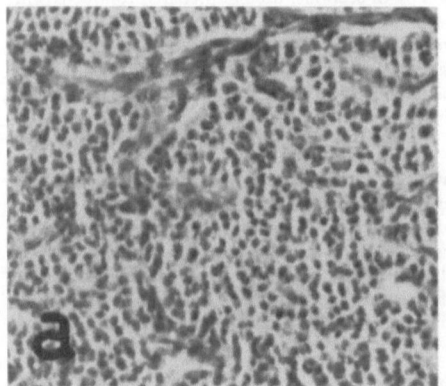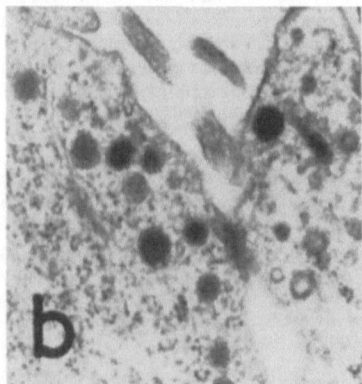

Fig. 2.15a,b. Bronchial carcinoid tumour. **a** Medium-power light microscopy shows a uniformly cellular tumour lacking recognisable differentiated characteristics. The stroma is richly vascular. Haematoxylin and eosin, × 215. **b** Electron microscopy shows numerous round electron-dense intracytoplasmic APUD granules. (The argyrophil reaction of this tumour is shown in Fig. 3.13.) × 25 000

These four examples show that the correlation between magnification and information yield is often negative; diverticular disease cannot be diagnosed by electron microscopy and APUD granules are invisible to the naked eye. Certainly, it is not worth doing electron microscopy unless one has clearly considered the indications (Chap. 9). Electron microscopy is not generally helpful in distinguishing a benign tumour from a malignant tumour or even from non-neoplastic tissue. It will, however, give useful information regarding histogenesis.

References

Abrahams C (1978) The "scrimp" technique—a method for the rapid diagnosis of surgical pathology specimens. Histopathology 2 : 255–266

Baron E (1939) Aspiration for the removal of biopsy material from the liver. Arch Intern Med 63 : 276–289

Baunsgaard P, Sanchez GC, Lundborg CJ (1979) The variation of pathological changes in the liver evaluated by double biopsies. Acta Pathol Microbiol Scand [A] 87 : 51–57

Berg JW, Robbins GF (1962) A late book at the safety of aspiration biopsy. Cancer 15 : 826–827

Berlin RB, Brines OA (1951) Detection of lymph node involvement in cancer. Am J Clin Pathol 21 : 332–337

Blaustein PA, Silverberg SG (1977) Rapid cytological examination of surgical specimens. Pathol Ann 2 : 251–278

Davies JD, Roberts G, Richardson PJ (1973) A serial whole-organ slicing technique for examining surgically resected breasts. J Clin Pathol 26 : 891–892

Dudgeon LS, Patrick C (1927) A new method for the rapid microscopical diagnosis of tumours with an account of 200 cases so examined. Br J Surg 15 : 250–260

Farrer-Brown G, Bennett MH, Harrison CV, Millett Y, Jelliffe AM (1972) The diagnosis of Hodgkin's disease in surgically excised spleens. J Clin Pathol 25 : 294–300

Franks LM (1954) Latent carcinoma of the prostate. J Pathol Bacteriol 68 : 603–616

Gallagher HS, Martin JE (1969) The study of mammary carcinoma by mammography and whole organ sectioning. Cancer 23 : 855–873

Godwin JT (1976) Rapid cytologic diagnosis of surgical specimens. Acta Cytol 20 : 111–115

Hatfield ARW, Slavin G, Segal AW, Levi AJ (1975) Importance of the site of endoscopic gastric biopsy in ulcerating lesions of the stomach. Gut 16 : 884–886

Hawk WA, Crile G, Hazard JB, Barrett DL (1966) Needle biopsy of the thyroid gland. Surg Gynecol Obstet 122 : 1053–1065

Kawai K, Takemoto T, Suzuki S, Ida K (1979) Proposed nomenclature and classification of the dye-spraying techniques in endoscopy. Endoscopy 1 : 23–25

Kellow WF, Cotsonas NJ, Chomet B, Zimmerman HJ (1959) Evaluation of the adequacy of needle biopsy specimens of the kidney: an autopsy study. Arch Intern Med 104 : 353–359

Kennedy A (1977) Basic techniques in diagnostic histopathology. Churchill Livingstone, London Edinburgh

Maharaj B, Maharaj RJ, Leary WP, Cooppan RM, Naran AD, Pirie D, Pudifin DJ (1986) Sampling variability and its influence on the diagnostic yield of percutaneous needle biopsy of the liver. Lancet I: 523–525

Meatheringham RE, Ackerman LV (1947) Aspiration biopsy of lymph nodes: a critical review of results of 300 aspirations. Surg Gynecol Obstet 84 : 1071–1976

Meyer JS (1973) Rapid cytologic identification of renal cortex in kidney biopsies. Am J Clin Pathol 59 : 595–600

Monroe CW (1948) Lymphatic spread of carcinoma of the breast. Arch Surg 57 : 479–486

Moore RD, Reagan JW (1953) Cellular study of lymph node imprints. Cancer 6 : 606–618.

Nichols TM, Boyes DA, Fidler HK (1968) Advantages of routine step serial sectioning of cervical cone biopsies. Am J Clin Pathol 49 : 342–346

Pease DC (1964) Histological techniques for electron microscopy. Academic Press, New York London, p 13

Pickren JW (1961) Significance of occult metastases: a study of breast cancer. Cancer 14 : 1266–1271

Pierson KK (1980) Principles of Prosection: a guide for the anatomic pathologist. Wiley, New York Chichester Brisbane Toronto

Rosai J (1981) Manual of surgical pathology and gross room procedures. University of Minnesota Press, Minneapolis

Rosenberg E, Black H (1959) The value of biopsy of cores of fresh hapatic tissue in the diagnosis of schistosomiasis. Am J Clin Pathol 32: 472–473

Rubio CA, Thomassen P, Sodeberg G, Kock Y (1978) Big cones and little cones. Histopathology 2: 133–143

Saltzstein SL (1960) Histologic diagnosis of breast carcinoma with the Silverman needle biopsy. Surgery 48: 366–374

Saphir O, Amromin GD (1948) Obscure axillary lymph-node metastasis in carcinoma of the breast. Cancer 1: 238–241

Silverberg SG, Vidone RA (1966) Carcinoma of the thyroid in surgical and post-mortem material. Analysis of 300 cases at autopsy and literature review. Ann Surg 164: 291–299

Soloway RD, Bagenstoss AH, Schoenfield LJ, Summerskill WHJ (1971) Observer error and sampling variability tested in evaluation of hepatitis and cirrhosis by liver biopsy. Am J Dig Dis 16: 1082–1086

Suen KC, Wood WS, Syed AA, Quenville NF, Clement PB (1978) Role of imprint cytology in intraoperative diagnosis: value and limitations. J Clin Pathol 31: 328–337

Te Velde J, Burkhardt R, Kleiverda K, Leenheers-Binnendijk L, Sommerfield W (1977) Methylmethacrylate as an embedding medium in histopathology. Histopathology 1: 319–330

Wagoner G, Ulevitch H, Gall EA, Schiff L (1951) Biopsy of needle specimens of liver tissue. VI. Comparison of findings on biopsy and at autopsy. Am J Clin Pathol 21: 338–341

Wilkinson EJ, Hause L (1974) Probability in lymph node sectioning. Cancer 33: 1269–1274

3   The Use of Stains

Little can be seen in unstained frozen or paraffin wax sections when they are examined by simple light microscopy. Phase-contrast, dark-ground, and interference microscopy will reveal more detail, but these methods, though they certainly do have applications, are unsuitable for most diagnostic needs. Fortunately a kaleidoscopic variety of stains is available for the study of sections of human tissues, ranging from general purpose stains, for just demonstrating the structure of cells and tissues, to stains which can detect individual substances with exquisite specificity.

One of the best ways for the trainee histopathologist to learn how to interpret the results of special stains is first to apply them to normal tissues and to typical lesions in which the diagnosis is not in any doubt. Special stains are rarely done on biopsies in which the nature of the disease is so obvious that the haematoxylin and eosin stain suffices for interpretation. With atypical lesions the features picked up by a special stain may be so subtle that one must know exactly what is to be looked for.

Another useful exercise for the trainee is to spend some time on the technical side of the laboratory. Histotechnological skills are best appreciated by trying to embed, cut and stain tissues oneself. The results may be of indifferent quality, but at least one learns to admire the expertise of those who do this every day. Some special stains are notoriously fickle; they demand considerable expertise for reliable results.

Principles of Staining

Stains may be either empirical or rational; some in the latter group were discovered empirically, but are now known to have a rational basis. Stains can be categorised according to the physico-chemical principles governing their action (Fig. 3.1).

1. There are stains which involve electrostatic attraction between oppositely charged dyes and tissue substances. Cationic tissue loci (e.g. those conferred by NH_3+ groups of basic amino acids) have an affinity for anionic dyes like eosin. Haematein (the oxidised form of haematoxylin), a cationic dye, is attracted to negatively charged groups in tissues.

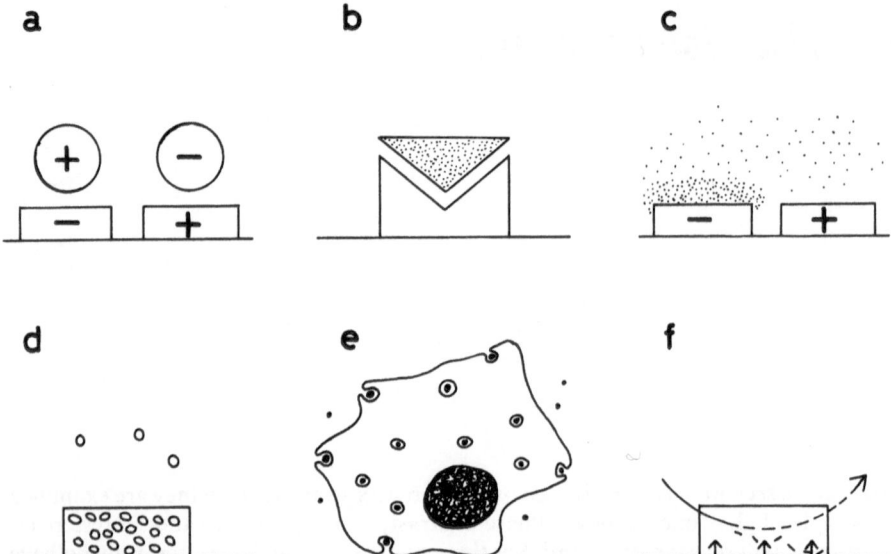

Fig. 3.1a–f. Principles of the major types of staining reaction. a Electrostatic attraction between cationic or anionic dyes and oppositely charged substances in sections. b Covalent binding between a dye and a reactive substance in the section. c Precipitation of a charged colloid over a section. d Solubility of a dye in the lipid phase of a section. e Vital staining—uptake of a dye by living cells. f Interaction of reactive groups in section with an added stain leading to a visible colour change (metachromasia) or precipitation.

2. Some stains result in the formation of a covalent bond between a dye, or chromogenic substance, and some substance in the tissue. For example, periodic acid oxidises carbohydrate radicals to aldehydes which then react with Schiff's reagent to give a reddish colour.

3. Charged colloids formed over sections will be attracted to oppositely charged tissue substances. This is the principle of Hale's colloidal iron method for some mucins.

4. Lipids can be stained by treating frozen sections with lipid-soluble dyes (e.g oil red O).

5. There are the vital stains pioneered by Paul Ehrlich in the 1850s. These can be subdivided into those that are supra-vital methods, done on living tissue outside the body, and intra-vital methods, done in the living body. An example would be Ehrlich's methylene blue technique for the demonstration of nerve endings.

6. Finally, there are histochemical and cytochemical methods in which substances, like enzymes, react with reagents applied to sections to give coloured or opaque products, sometimes as a precipitate. Examples include Perls' stain for haemosiderin, in which acid liberates ferric iron from haemosiderin and this reacts with potassium ferrocyanide to give Prussian blue (ferric ferrocyanide), and the argentaffin reaction, in which silver nitrate is reduced to metallic silver by the granules of melanocytes or Kultchitsky cells. Alternatively, the substance in the tissue may alter the conformational structure of the dye so that its colour is changed; this is called metachromasia. For example, toluidine blue is changed to purple by acid mucopolysaccharides. Unfor-

tunately, because fixation in formalin denatures potentially reactive substances and some water-soluble substances may be further lost in processing, frozen sections often have to be used for enzyme histochemistry. Indeed, although most stains are devised to work well or even optimally after formalin fixation of the tissue, the fixative should be selected bearing in mind the tinctorial methods that are likely to be applied.

Most stains include a counterstain (e.g. neutral red in Perls' stain) so that positive cells, granules, fibres or organisms stand out against a contrasting background.

Positive control sections are important, particularly if the stain is only infrequently done.

Indications for Special Stains

The designation of a stain as "special" may be arbitrary, but for these purposes I will regard a special stain as one other than haematoxylin and eosin, Papanicolaou or Giemsa; these are the commonly used general purpose stains for sections, smears and imprints.

Special stains should not be used just for their aesthetic appeal or as a substitute for seeking additional clinical information. Nor should special stains be requested as a ruse to allow one to ponder over the haematoxylin and eosin stained sections a little longer! One can also become so engrossed with the elegance of some stains that important clinical features are ignored. More adequate sampling of the specimen may be more useful than special stains done on an initial limited tissue sample.

General guidance about the use of special stains is difficult to give because each pathologist has an individual way of working; the following suggestions are unlikely to be contentious, complete, or necessary in every case.

Inflammatory lesions—stain for organisms likely to excite the type of reaction that is seen in the tissue, e.g. pyogenic reaction—Gram's stain for most bacteria; granulomatous reaction—Ziehl-Neelsen for *Mycobacterium tuberculosis*, Wade-Fite for *Mycobacterium leprae*, periodic acid–Schiff and methenamine (hexamine) silver for fungi.

Tumours—stain for evidence of differentiation if this is not evident in haematoxylin and eosin stained sections, e.g. adenocarcinoma—alcian blue/periodic acid-Schiff (PAS) and mucicarmine for mucin; carcinoid—Masson-Fontana argentaffin and diazo stain for 5-hydroxytryptamine-rich granules.

Lymph node biopsies—silver impregnation method for reticulin fibres to assess general architecture; periodic acid-Schiff method for histiocytes, immunoglobulin, vascular basement membrane; methyl green-pyronin (Unna-Pappenheim) for plasma cells and immunoblasts; immunohistology for lambda and kappa light chains to help distinguish between monoclonal (neoplastic) and polyclonal (reactive) B-cell proliferation, and for lymphocyte surface markers.

Liver biopsies—stain reticulin for general architecture; Masson trichrome also for architectural integrity and evidence of fibrosis; Shikata orcein method or aldehyde fuchsin for HB_sAg, elastic fibres and copper-binding protein; periodic acid-Schiff stain after diastase treatment of section for retention globules in a-1-antitrypsin deficiency; Perls' stain for haemosiderin; rubeanic acid or rhodanine for copper.

Kidney biopsies—periodic acid-Schiff and reticulin stain for basement membrane configuration; Martius yellow-scarlet-blue (MSB) for fibrin; Masson trichrome for interstitial fibrosis; immunohistology for immune complexes, complement, etc.

Gastrointestinal biopsies—alcian blue/periodic acid-Schiff stain for identification of metaplasia, mucin depletion, muciphages, etc.

Muscle biopsies—enzyme histochemistry (e.g. succinic dehydrogenase, ATPase) for fibre types; methylene blue for motor end plates, etc.

Bone marrow—reticulin stain for fibrosis; Perls' method for haemosiderin; methyl green-pyronin for plasma cells and immunoblasts; Leder's stain for granulocyte precursors.

Identification of Specific Substances and Cells

This section briefly describes the use of stains to demonstrate specific components of tissues and summarises their main diagnostic value. More detailed information about various stains, particularly of a technical nature, can be found elsewhere (Bancroft and Stevens 1975). Some of the more commonly used special stains are listed in Table 3.1.

Table 3.1. Some commonly used special stains

Stain	Substance	Colour
ARDY[a]	Muscle	Red
Alcian blue	Acid mucins	Blue
Aldehyde fuchsin	Elastin	Purple
	HBsAg	
	Cu-binding protein	
	Mast cells	
	β-cell granules	
Alizarin red	Calcium salts	Red
Pascual	Argyrophil granules	Black
Congo red) } Sirius red) }	Amyloid	Red
Diazo	5HT	Orange/brown
Dieterle	*L. pneumophila*	Black
van Gieson[b]	Collagen	Red
Gram[c]	Micro-organisms	Blue/red
Grimelius	Argyrophil granules	Black
Hale's colloidal iron	Acid mucins	Blue
Holmes'	Nerve fibres	Black
von Kossa[d]	Phosphate	Black
Leder	Mast cells, myeloid series, e.g. granulocytes	Red
Luxol fast blue	Myelin	Blue
MSB	Fibrin	Red
Masson-Fontana	Melanin	Black
	Argentaffin granules	
Masson trichrome	Collagen	Green
	Muscle, hepatocytes, etc.	Red
Silver methenamine[e]	Basement membrane	
	Fungi	Black

Table 3.1 *(continued)*

Stain	Substance	Colour
Methyl green/pyronin	DNA	Green
	RNA	Red
Miller's elastin[f]	Elastin	Black
Mucicarmine	Mucins	Red
Oil red O[g]	Lipids	Red
Orcein	Elastin	Brown
	HBsAg	
	Cu-binding protein	
Palmgren	Nerve fibres	Black
PAS	Glycogen	Purple
	Mucins	
	Fungi	
Perls'	Haemosiderin	Blue
Phloxine-tartrazine	Viral inclusions, keratin, etc.	Red
PTAH	Muscle	Blue
Reticulin[h]	Reticulin fibres	Black
Rhodanine	Copper	Red
Rubeanic acid	Copper	Green/black
Solochrome cyanin	Myelin	Blue
Thioflavine-T	Amyloid	Fluorescent
Wade-Fite	Weakly acid-fast bacilli, e.g. *M. leprae*	Red
Warthin-Starry	Spirochaetes	Black
Ziehl-Neelsen	Acid-alcohol fast bacilli; lipofuscin; *S. mansoni*	Red

[a] See text for explanation of abbreviations used in this table.
[b] Usually combined with elastin stain.
[c] Several variants for histology.
[d] Usually used to detect calcification invariably associated with phosphate.
[e] Hexamine (U.S.A.).
[f] Several alternatives, e.g. Weigert's method.
[g] Requires frozen sections.
[h] Silver impregnation technique—some variants.

Connective Tissues

Collagen, including reticulin (type III collagen), is the major structural protein of tissues providing mechanical support to parenchymal elements. Parenchymal distortion, such as that seen in hepatic fibrosis or cirrhosis, is more easily seen in sections stained for collagen: van Gieson's stain, Masson's trichrome and silver impregnation techniques are all reliable methods (Fig. 3.2). Other than reticulin, the various collagen types cannot be distinguished by simple tinctorial methods, but it is possible to do so by immunological staining methods.

The reticulin framework of tumours is sometimes used as a guide to histogenetic differentiation; carcinomas tend to have cells arranged in groups invested by a fairly coarse reticulin mesh, whereas most lymphomas have a more delicate reticulin pattern with less "boxing" of the cells. In necrotic tissue, including tumours, the reticulin framework persists long after cytological detail has been lost; this can be used to assess the architecture of lesions in biopsies which are substantially necrotic.

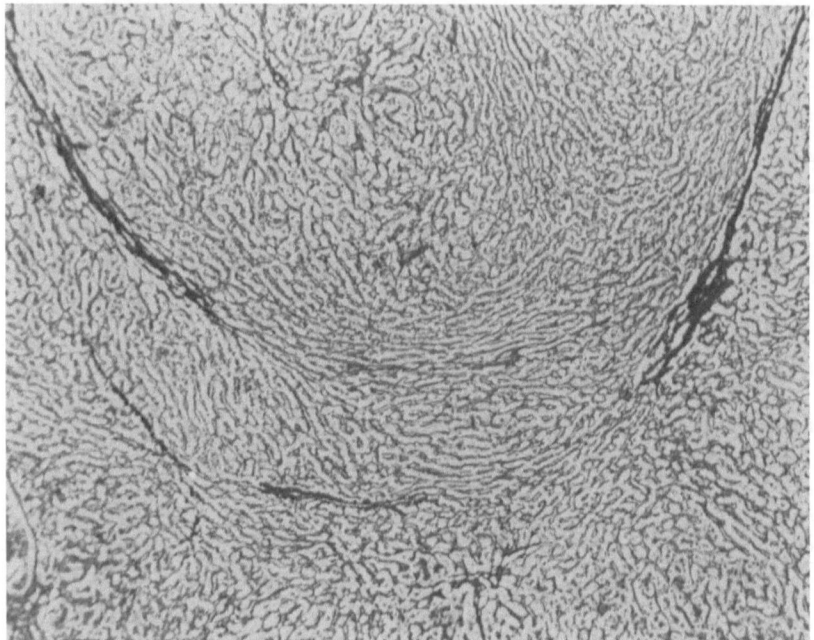

Fig. 3.2. Nodular regeneration in macronodular cirrhosis emphasised by deformation of the reticulin framework. Wedge biopsy. Gomori's method for reticulin, × 56

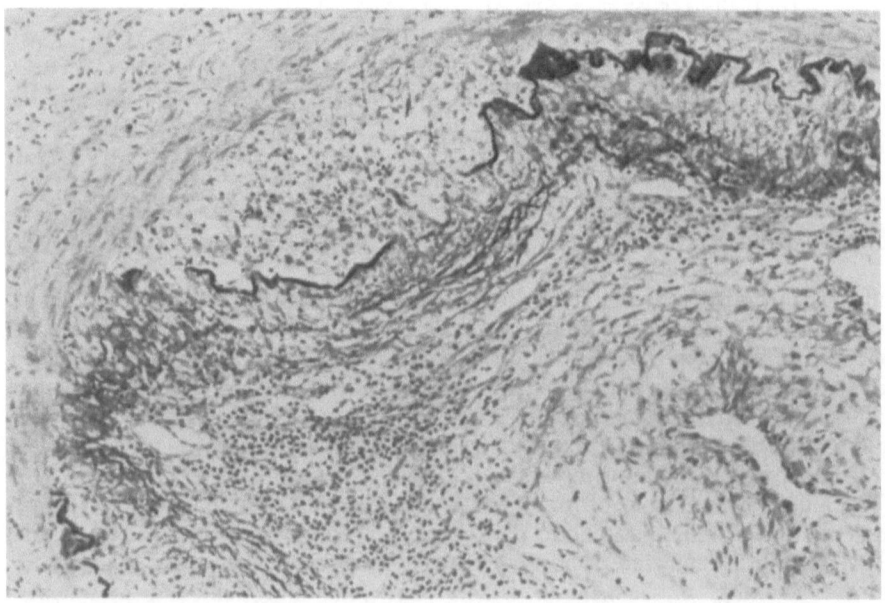

Fig. 3.3. Temporal artery biopsy showing disruption of the internal elastic lamina, active inflammation,
and intimal proliferation in giant-cell (temporal) arteritis. Miller's elastic/van Gieson, × 215

Elastin can be identified using orcein, aldehyde fuchsin, Weigert's or Miller's methods. Additionally, it shows bright autofluorescence by ultraviolet microscopy. Stains for elastin are indispensable in the examination of vascular lesions, either to identify damaged vessels showing disruption of elastic laminae or to define the exact location of a lesion in the vessel wall (Fig. 3.3). Elastin stains are also useful in the diagnosis of skin biopsies and lung biopsies; elastin fibres are a major structural component of these tissues. Vascular invasion by tumours is easier to find in sections stained for elastin because vessels totally occluded by plugs of tumour can still be recognised from surviving concentric elastic laminae (Fig. 3.4).

Basement membranes contain a glycoprotein component and are thus periodic acid-Schiff (PAS) positive.

Normal muscle cells can usually be identified without difficulty, but muscle-cell tumours may give problems. Phosphotungstic acid-haematoxylin (PTAH) and acid red/direct yellow (ARDY) are fairly reliable stains for muscle cells and, if properly done, myofibrils or striations should be evident (Fig. 3.5). Muscle striations are also birefringent.

Fibrin may show variable staining reactions according to the degree of polymerisation. PTAH and MSB stains are both useful; the latter gives particularly vivid results (fibrin is stained red).

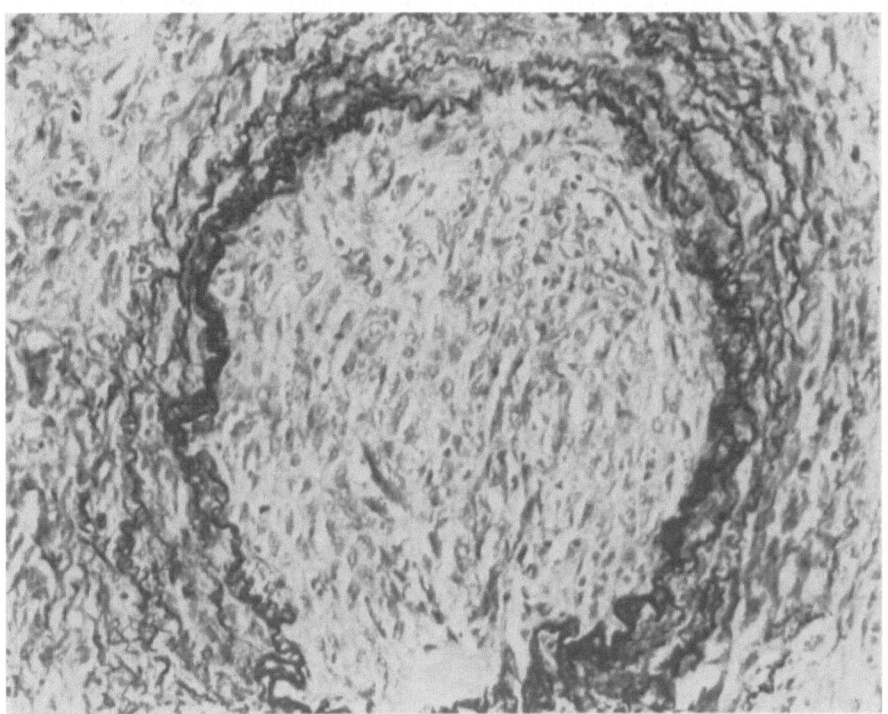

Fig. 3.4. Identification of vascular invasion, by an osteosarcoma, assisted by delineation of the elastic laminae in the vessel wall. Vessels obliterated by tumour may be difficult to recognise in sections routinely stained with haematoxylin and eosin. Miller's elastic/van Gieson, × 215

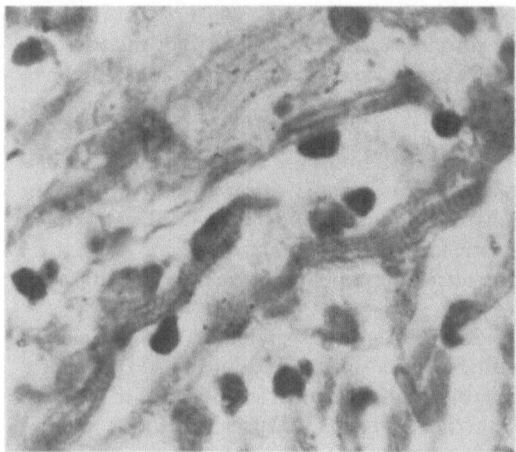

Fig. 3.5. Muscle striations in an embryonal (botryoid) rhabdomyosarcoma. Phosphotungstic acid–haematoxylin, × 1050

Amyloid substances are glycoproteins with a β-pleated sheet configuration; this confers certain tinctorial properties on the material, such as an affinity for Congo red, Sirius red and thioflavine T (Fig. 3.6). The specificity of Congo red or Sirius red staining can be enhanced by examining the section through crossed polarising filters; as one filter is rotated almost to the point of extinction the colour shifts from red to green, a phenomenon known as dichroism. Further characterisation of the

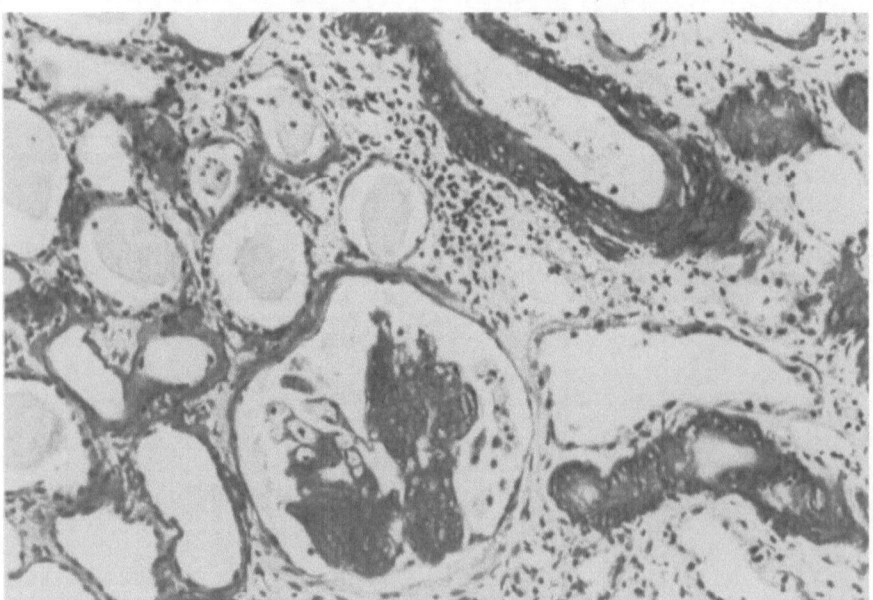

Fig. 3.6. Deposition of an amyloid substance in the kidney; renal biopsy from a patient with nephrotic syndrome. The substance is present in the glomerular tuft, tubular basement membrane, and blood vessel walls. Sirius red, × 130

amyloid substance requires other methods. The amyloid substances of primary amyloidosis consist of immunoglobulin light chains, at least in part, and can be identified by immunofluorescence or immunohistochemistry using antisera to kappa and lambda chains and amyloid A protein. The affinity of primary and multiple myeloma amyloid for Congo or Sirius red is resistant to pretreatment of the tissue sections with potassium permanganate. Secondary amyloid is permanganate sensitive (Wright et al. 1977). APUD amyloid found in the stroma of, for instance, medullary carcinoma of the thyroid has some amino acid sequences common to calcitonin. It lacks tryptophan and tyrosine and therefore gives negative reactions with dimethylaminobenzaldehyde (Pearse et al. 1972).

Most central and peripheral nervous system tumours can be diagnosed on good haematoxylin and eosin stained sections. Some gold and silver impregnation techniques exist for the demonstration of glial cells and fibres (e.g. Cajal, Hortega methods), but these need care for reliable results. PTAH is an easier method for demonstrating astrocytic fibres. Nerve fibres can be stained by the Holmes silver impregnation technique. Myelin can be stained with luxol fast blue. Degenerate myelin releases neutral fats and fatty acids which, in frozen sections or freshly teased nerves, can be picked up with lipid stains like oil red O.

Carbohydrates

Carbohydrates of diagnostic importance can be divided into glycogen and mucins.

Glycogen is a polysaccharide. The simplest specific method for its detection is the PAS technique; preincubation of the section with diastase will specifically abrogate staining due to glycogen. Unreliable results may be obtained if the tissue is not fixed with some care because glycogen is fairly water soluble. Glycogen storage diseases require histochemical methods for precise diagnosis (Lake 1970).

Mucins are polysaccharides covalently linked to a variable protein moiety. Though their classification is complex, and often inconsistent, they can be divided into two major groups—neutral and acidic. Neutral mucins include those of the gastric mucosa, colorectal mucosa and Brunner's glands; they are PAS positive. Acid mucins are classified as follows:

1. Sulphated
   a) Strongly sulphated mucins of connective tissues—alcian blue (AB) positive, PAS negative
   b) Weakly sulphated mucins of epithelia (salivary glands, duodenal and colonic goblet cells)—AB positive, PAS variable
2. Non-sulphated
   a) Sialic acid rich (goblet cells of lung and intestine)—AB and PAS positive
   b) Hyaluronic acid rich (synovium, mesothelia, and skin)—AB positive, PAS negative

Alcian blue binds to mucins through electrostatic forces, an interaction which is pH dependent; at pH 0.5 it will stain sulphated connective tissue mucins only, whereas at pH 2.5 it will bind to sulphated epithelial mucins and non-sulphated mucins. Above pH 3.0 it loses its specificity for acid mucins. Schiff's reagent reacts with aldehydes generated in carbohydrate moieties through oxidation by periodic acid. A justifiably

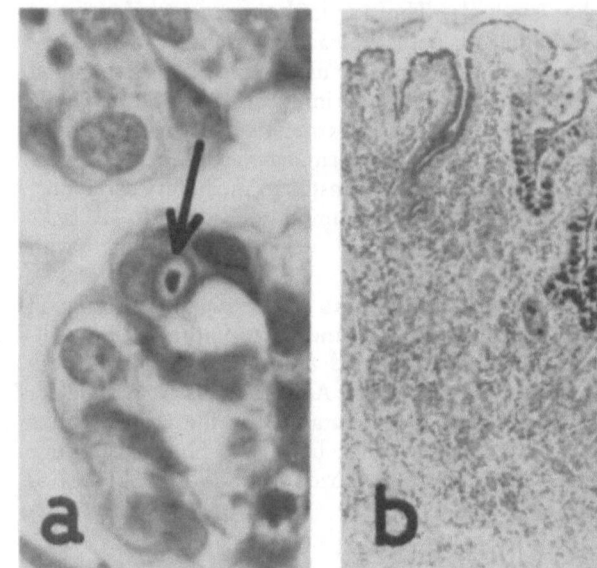

Fig. 3.7a,b. Diagnostic applications of mucin stains. **a** Mucin-containing vacuole ("private acinus") in lobular carcinoma-in-situ of breast. Alcian blue/periodic acid-Schiff, × 1344. **b** Intestinal metaplasia (*right*) in the stomach. Alcian blue/periodic acid-Schiff, × 65

popular broad-spectrum mucin stain is the AB/PAS combination. Alcian blue is applied to the section first and blocks any mucins that might also react with periodic acid-Schiff reagent. Southgate's mucicarmine can also be used, but may not stain strongly acidic sulphated mucins.

The specificity of mucin stains can be increased by testing the ability of mucin-degrading enzymes to abolish staining; pretreatment of sections with neuraminidase should block sialomucin staining and hyaluronidase pretreatment eliminates staining due to hyaluronic acid-rich mucins.

A major use of mucin stains is the investigation of metastatic malignancy. The presence of mucin within the tumour is an indication of glandular differentiation (i.e. adenocarcinoma, Fig. 3.7) and the exact type of mucin may point towards a particular primary origin (Cook 1973). It is claimed that mesotheliomas, which sometimes mimic adenocarcinomas in their growth pattern, can be identified by virtue of alcian blue staining sensitive to hyaluronidase pretreatment (Wagner et al. 1962). The O-acylated sialic acids in epithelial mucin of the lower gastrointestinal tract are specifically stained by the periodate-borohydride/KOH/PAS technique (Culling et al. 1975); this can be used to help distinguish metastatic colorectal adenocarcinomas from other primary and secondary adenocarcinomas.

Nucleic Acids

There is usually little difficulty in identifying the nucleus within a cell, so stains for DNA alone find little place in diagnostic histopathology. RNA is not so readily detected. Unna and Pappenheim are credited with the development of the methyl green/

pyronin stain for nucleic acids; DNA takes up methyl green and RNA is pyroninophilic. This is a useful method for picking up plasma cells in inflammatory lesions or lymph nodes and as an aid to the identification of myeloma; the cytoplasm of immunoblasts is pyroninophilic.

Fats

Fat dissolves in alcohol and other solvents used in processing tissue for wax embedding; frozen sections of fresh or formalin-fixed tissue must be used. Oil red O is a widely used stain, but there are others.

Common instances where the demonstration of fat is useful include specimens in which fat embolism is queried, xanthomatous lesions, and suspected primary tumours of adipose tissue.

Minerals and Pigments

The detection of specific elements in tissue is best done by neutron activation analysis or atomic absorption spectrophotometry. However, these methods give no information about the localisation of the elements. X-ray microanalysis may be used at the ultrastructural level, but there are problems with sampling. Few diagnostic laboratories have access to these facilities.

Some tinctorial methods are available for a few of the biologically important metals. These include Perls' stain for iron as haemosiderin and the rubeanic acid or rhodanine methods for copper (Fig. 3.8). Orcein also stains the sulphydryl-rich copper-associated protein seen in the periportal hepatocytes of patients with chronic cholestasis.

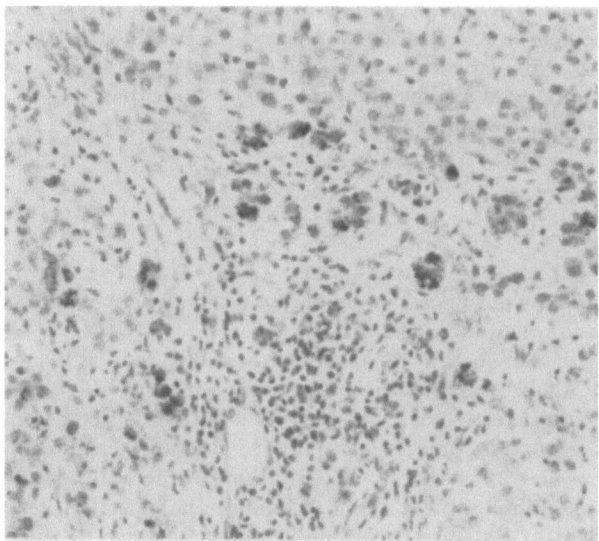

Fig. 3.8. Granular cytoplasmic staining of periportal hepatocytes by the rubeanic acid method for copper. Needle liver biopsy from a patient with primary biliary cirrhosis. Rubeanic acid, × 558

Fig. 3.9. Tetracycline labelling of bone. Two spaced doses of tetracyline were given before the bone biopsy. Examination of the section with ultraviolet light reveals two mineralisation fronts, labelled by the fluorescent tetracyline, encircling an Haversian canal. The dynamics of bone formation can be examined in this way, × 220

Bone mineral must be assessed in undecalcified specimens; this needs special microtomy techniques for section preparation. Von Kossa's method demonstrates the phosphate that is usually associated with calcium deposition. Alizarin red S stains calcium itself. Tetracyline labelling in vivo is a useful method for studying the rate of osteogenesis, the band of tetracyline-labelled bone is seen as a fluorescent line close to the edge of trabeculae in sections examined by ultraviolet microscopy (Fig. 3.9).

The identification of pigments and crystals is further discussed in Chap. 5.

Organisms

Histology is not the ideal method for the identification of organisms; where possible, fresh tissue should be obtained for microbiological studies.

Viruses can be seen only with the electron microscope, though the much larger inclusion bodies may betray their presence in histological sections. Phloxine/ tartrazine is a useful method for the detection of inclusions, but it must be remembered that tissues contain other phloxinophilic objects (e.g. keratin). The surface antigen of hepatitis virus B (HB_sAg) resides in the liver-cell cytoplasm of infected carriers and can be detected there by immunological staining techniques. However, HB_sAg is rich in sulphydryl groups and can be stained by the Shikata orcein method or aldehyde fuchsin (Fig. 3.10). Positive cells often have a "ground glass" cytoplasm in haematoxylin and eosin preparations.

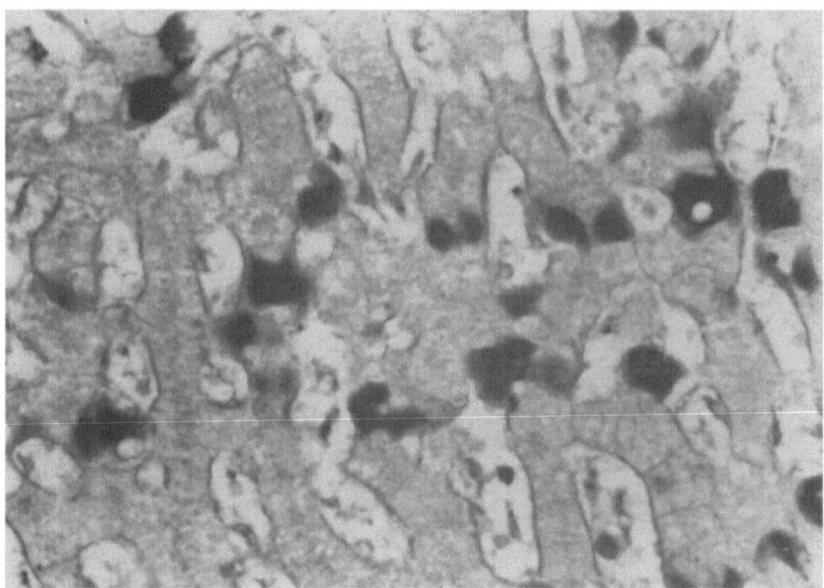

Fig. 3.10. Hepatocytes laden with intracytoplasmic hepatitis B surface antigen. Aldehyde fuchsin, × 489

Gram's stain, or modifications of it, is suitable for most bacteria, but Gram-negative organisms are often difficult to see in sections. The green background of the Gram-Twort method may allow Gram-negative organisms to stand out more clearly. Dieterle's stain must be used for the otherwise tinctorially elusive *Legionella pneumophila*. Acid-fast myobacteria (e.g. *Mycobacterium tuberculosis*) are stained by the Ziehl-Neelsen method or the fluorescent auramine T method. *Mycobacterium leprae* will be missed unless a Wade-Fite stain is done. Decalcification of a specimen (e.g. bone biopsy) in agents containing greater than 2.5 M hydrochloric acid (e.g. RDC, Bethlehem Instruments Ltd.) may lead to false-negative Ziehl-Neelsen staining for acid-fast bacilli (Anderson and Coup 1975).

Metazoan parasites, such as worms and flukes, are easy to see though sometimes less easy to identify in sections stained with haematoxylin and eosin. However, protozoa can be missed, not only because they are small but also because they may be dismissed as degenerate cells or mere flecks of mucus. The PAS method is recommended for the detection of amoebae in colorectal biopsies. *Leishmania* organisms stain up quite well in smears or sections stained by the Giemsa method. *Giardia lamblia* in jejunal biopsies is weakly haematoxyphilic, but stains more intensely with PAS or PTAH. The ova of *Schistosoma mansoni* are acid-fast by Ziehl-Neelsen's method.

Fungi have a glycoprotein wall; this stains intensely by the PAS method. Methenamine (hexamine) silver is also recommended (Fig. 3.11). The mucinous capsule of *Cryptococcus neoformans* stains strongly with mucicarmine.

Spirochaetes are best sought for in tissue blocks stained by Levaditi's method, or by employing the Warthin-Starry stain which is designed for use on sections.

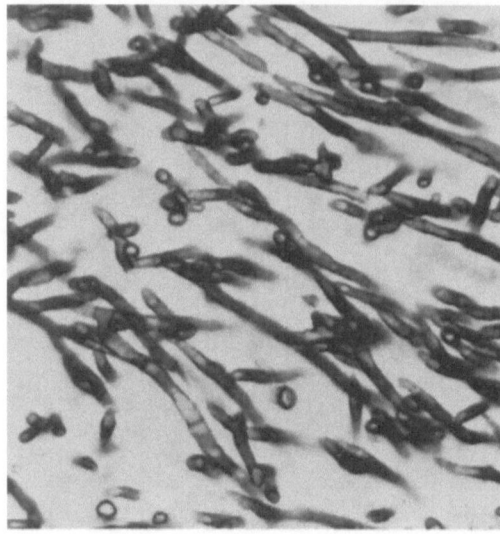

Fig. 3.11. Fungal hyphae, probably *Aspergillus* sp., in a pulmonary mycetoma. The walls of the hyphae are strongly stained and, in a few places, septae can be discerned. Methenamine (hexamine) silver, × 558

Specific Cells

Cells are identified by the presence of differentiated features. Sometimes these features are obvious, such as with muscle striations or a keratinising squamous cell. On many occasions, particularly with tumour diagnosis, special stains may be needed to dem-

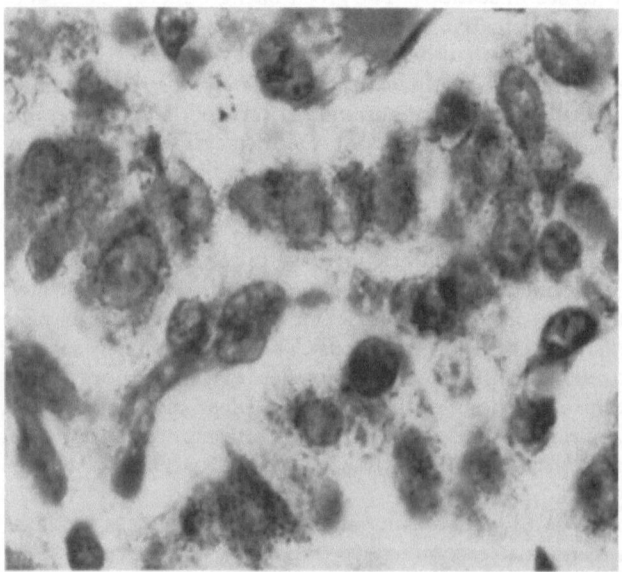

Fig. 3.12. Bronchial carcinoid tumour showing a positive argyrophil reaction. There are numerous fine intracytoplasmic granules. The argentaffin reaction was negative. (The ultrastructural appearances are shown in Fig. 2.15.) Pascual argyrophil method, × 1050

onstrate differentiated features that are usually quite obvious in non-neoplastic cells of the same lineage stained by haematoxylin and eosin.

Endocrine cells merit detailed consideration. The most specific identification is achieved by actually demonstrating synthesis of the hormone in question, but this is not possible with fixed histological preparations. Almost as specific is the demonstration of the hormone in the cell, on the assumption that the cell has actually synthesised it. Bioassay is obviously impractical and so one has to resort to immunological methods with specific antisera to hormones; peptide hormones can be successfully demonstrated in this way (see p. 66).

Either argyrophilia or a positive argentaffin reaction characterise the cell as one of the APUD series. Some APUD cells have the ability to reduce silver salts and trap the precipitated silver, i.e. argentaffin. Others are unable to reduce silver salts, but will trap silver precipitated in their vicinity, i.e. argyrophil (Fig. 3.12). Many argentaffin cells also give a positive diazo reaction. Additionally, cells that contain aromatic amine-rich granules (e.g. 5-hydroxytryptamine, histamine, catechol amines, dopamine) will exhibit formaldehyde-induced fluorescence.

Chromaffin cells, as of a phaeochromocytoma, have an affinity for chromate salts. When fixed in a dichromate solution such as Müller's fixative, the intracytoplasmic granules stain a green colour with Giemsa.

Most cells can be demonstrated in sections by aldehyde fuchsin, toluidine blue, Leder's stain, or Giemsa stain; in each case the cytoplasmic granules stain strongly. Giemsa stain is a good general purpose stain for all blood cells and cells of the blood-forming and lymphoid organs, and permits direct comparison between tissue imprints, sections, smears and blood films from the same patient.

Immunological staining methods can be used to detect immunoglobulins in or on lymphoid cells. The demonstration of light-chain restriction in lymphoproliferative lesions points to neoplasia rather than reactive conditions. Surface markers are best assessed in frozen sections (see p. 63).

Histochemical methods have been suggested for the accurate identification of lymphoid and haemopoietic neoplasms (Lennert 1978). These include Leder's chloroacetate esterase method for cells of the myeloid series (this can be done on paraffin wax sections), a-naphthyl acetate esterase for cells of histiocytic lineage, and tartrate-resistant acid phosphatase for hairy-cell leukaemia.

There are many other cell products for which no tinctorial method exists. It is likely that immunohistological methods will be increasingly used, and become more specific, reliable and sensitive.

Autoradiography

Autoradiography is a method for localising radioactive emission in tissue sections. The section is covered with a thin photographic emulsion, exposed in the dark through contact with the section, and developed photographically (Fig. 3.13). a-emission is visible as linear tracks of silver grains and β-emission as a fine deposit of silver grains over the source. γ-emission is difficult to localise precisely, but can be roughly done by leaving the section in contact with X-ray film.

Radioactive isotopes may be introduced into the tissue in vivo through investigative or therapeutic procedures (e.g. thorotrast, ^{131}I), or by incubating cells or fresh tissue

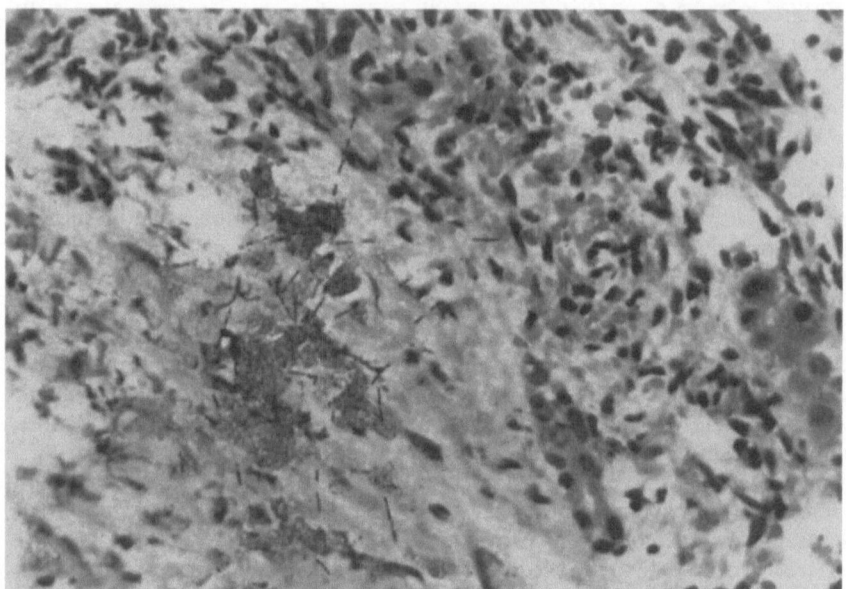

Fig. 3.13. Thorotrast particles in a needle liver biopsy identified by autoradiography. The linear tracks denote *a*-emission. The adjacent liver tissue contains an angiosarcoma presenting 36 years after cerebral angiography with thorotrast. Haematoxylin and eosin/autoradiograph, × 320

slices with labelled precursors in vitro. For example, incubation of fresh tissue with tritiated thymidine will label the nuclei of cells in DNA synthesis.

Autoradiography is a powerful research tool, but finds little application in routine diagnostic work.

References

Anderson G, Coup AJ (1975) Effect of decalcifying agents on the staining of *Mycobacterium tuberculosis*. J Clin Pathol 28: 774–745

Bancroft JD, Stevens A (1975) Histopathological stains and their diagnostic uses. Churchill Livingstone, Edinburgh London New York

Cook HC (1973) A histochemical characterisation of malignant tumour mucins as a possible aid in the identification of metastatic deposits. Med Lab Technol 30: 217–224

Cullin CFA, Reid PE, Burton JD, Dunn WL (1975) A histochemical method of differentiating lower gastrointestinal tract mucin from other mucins in primary or metastatic tumours. J Clin Pathol 28: 656–658

Lake BD (1970) The histochemical evaluation of glycogen storage diseases: A review of techniques and their limitations. Histochem J 2: 441–450

Lennert K (1978) Malignant lymphomas. Springer, Berlin Heidelberg New York, pp 73–82

Pearse AGE, Ewen SWB, Polak JM (1972) The genesis of apudamyloid in endocrine polypeptide tumours: histochemical distinction from immunoamyloid. Virchows Arch (Cell Pathol) 10: 93–107

Wagner JC, Munday DE, Harington JS (1962) Histochemical demonstration of hyaluronic acid in pleural mesotheliomas. J Pathol Bacteriol 84: 73–78

Wright JR, Calkins E, Humphrey RL (1977) Potassium permanganate reaction in amyloidosis: a histologic method to assist in differentiating forms of this disease. Lab Invest 36: 274–281

4　Immunohistology

The horizons of histopathology have been extended dramatically by the application of immunological methods to the detection and localisation of a wide range of otherwise invisible substances in cells and tissue sections. Although immunofluorescence has been available since the 1940s (Coons et al. 1941), this technique has been difficult to employ routinely in clinical histopathology because of the almost universal requirement for frozen sections of fresh tissue for optimum results. The recent resurgence of interest in immunohistology is attributable to two factors: first, the availability of antibodies conjugated to tracer substances, such as enzymes (Nakane and Pierce 1966), which can be made visible for conventional light microscopy (Table 4.1); second, the more recent use of monoclonal antibody technology (Kohler and Milstein 1975) to generate exquisitely specific reagents for histological applications.

Table 4.1. Relative merits of fluorescent and non-fluorescent tracers for immunohistology

	Tracer	
	Fluorescent[a]	Non-fluorescent[b]
Speed	Relatively rapid	Slower
Cost	Relatively cheap	More costly
Effect of fixation	Adverse (autofluorescence)	"Masking" of antigens can be reversed
Microscope	UV microscope	Conventional light microscope
Permanence	Preparations fade	Permanent
Counterstains	Limited	Wide range
Sections	Frozen sections usually essential	Sections from stored wax blocks can be used
Retrospective studies	Difficult	Easy

[a] e.g. fluorescein
[b] e.g. oxidised DAB as in the immunoperoxidase method

The immunological principles of immunofluorescence and the newer immunological tracing methods are the same. All these techniques require a primary antibody of defined specificity which binds to the substance (operationally behaving as the antigen) in the cell or tissue section. It is important to ensure that the immunoreactive portion of the substance (e.g. a hormone) corresponds as closely as possible to the portion that has specific biological activity, otherwise the antibody may cross-react with antigenic determinants resident on biologically unrelated substances. The primary antibody may be linked to a tracer or, more often, a secondary antibody linked to a tracer is used.

Techniques for Immunohistology

Because immunofluorescence has been largely displaced by the newer immunohistological methods employing tracers which are either innately visible or can be rendered visible for conventional light microscopy, we shall concentrate on these latter techniques for detailed consideration (Taylor 1978; Heyderman 1979). The techniques are summarised diagrammatically in Fig. 4.1.

Direct Method

This is the simplest method; a one-step procedure for which the antibody must first have been conjugated to a tracer (the choice of tracer is considered on p. 59. After preparing the histological section(s) in the way that is described on pp. 61–62, the appropriately diluted antibody-tracer conjugate is applied and then washed off after a predetermined optimum incubation time. This method is simple, quick and relatively inexpensive, but it suffers from inherently poor sensitivity because no amplification is involved; each epitope on the tissue substance being studied ultimately binds just one antibody molecule which in turn is capable of carrying a restricted number of tracer molecules. Because of poor sensitivity this method is rarely used for diagnostic work despite its speed and economy.

Indirect Method

This is a two-stage procedure (Heyderman and Neville 1977). First, the section or cell preparation is incubated with the primary antibody, optimally diluted, specific for the substance in question. After washing, the section is then exposed to a second antibody which has been conjugated to a tracer substance. This second antibody must have been raised in another species and should recognise the Fc portion of the immunoglobulin class and species of the primary antibody. For example, if the primary antibody is of IgG class and was raised in a rabbit, then the second antibody is raised by injecting rabbit IgG into another species such as a goat; the second antibody reacts with an epitope on the Fc portion of the primary antibody. This method is more time-consuming and slightly more costly, but the results are usually cleaner with less background staining.

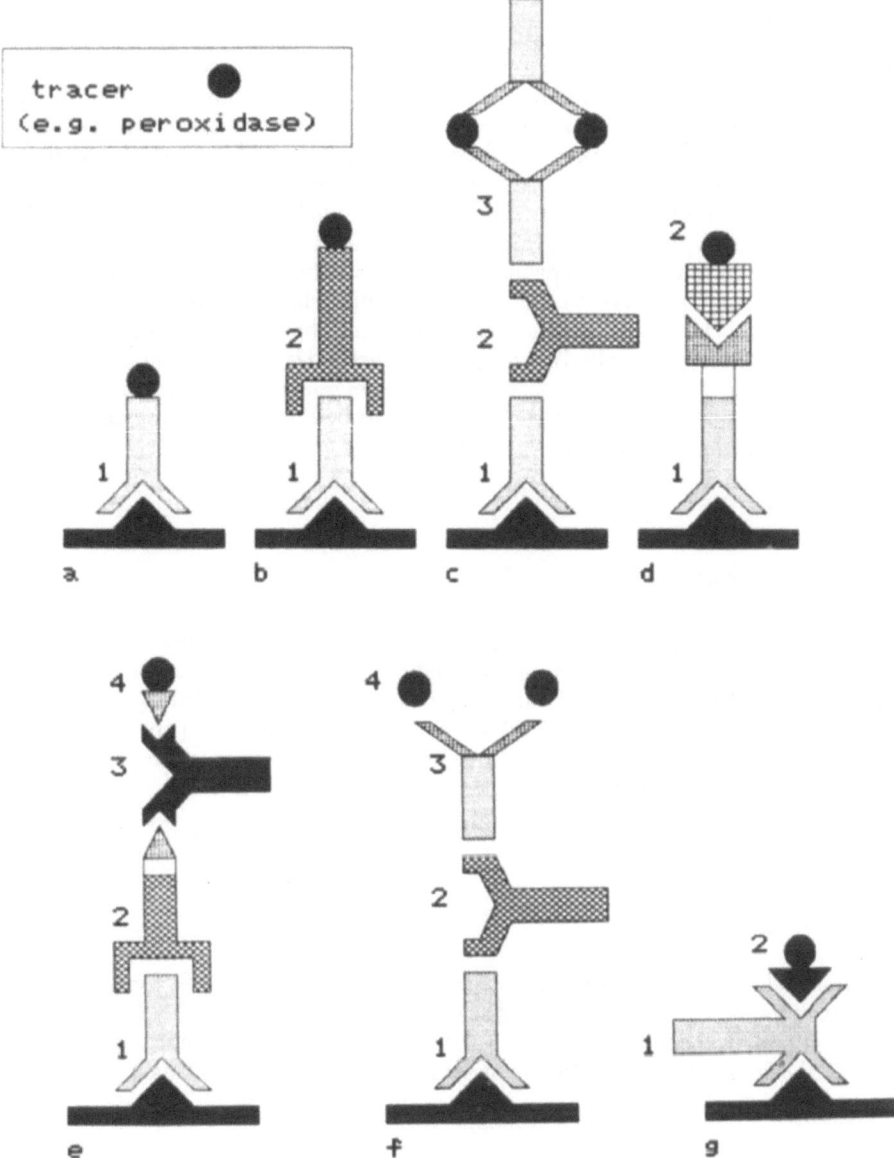

Fig. 4.1a–g. Diagrammatic summary of the various methods used in immunohistology. a Direct method: antibody-tracer conjugate (*1*). b Indirect method: primary antibody (*1*) is followed by second antibody-tracer conjugate (*2*). c Tracer-antitracer complex method: primary (*1*) and bridging antibody (*2*) are followed by tracer-antitracer immune complex (*3*). d Biotin-avidin method: primary antibody conjugated to biotin (*1*) is detected by tracer conjugated to avidin (*2*). e Hapten method: primary antibody (*1*) is traced by second antibody conjugated to hapten (*2*); anti-hapten antibody (*3*) bridges between second antibody-hapten conjugate and hapten-tracer conjugate. f Enzyme-bridge method: bridging antibody (*2*) links primary antibody (*1*) with the antitracer antibody (*3*), which is in turn located by exposure to the tracer (*4*). g Labelled-antigen method: primary antibody (*1*) binds to antigen in section and also to added antigen (*2*) which has been conjugated to a tracer.

Tracer-Complex Method

This requires the preparation of an antigen-antibody complex to be used in the final step (Sternberger et al. 1970). The complex consists of the tracer bound to an antibody that has been raised to it in the same species that was used to generate the primary antibody (e.g. peroxidase-antiperoxidase complex is produced by mixing peroxidase with an antibody that has been raised to it). For example, if the primary antibody has been raised in a rabbit, then the tracer-complex should be formed with an antibody that has been raised by injecting the tracer into a rabbit. Having prepared the complex, the section is first treated with the primary antibody, washed, incubated with a second unconjugated antibody raised to the immunoglobulin class and species of the first antibody (other "bridging" reagents are described below), washed, and then incubated with the tracer-antibody complex. This method is the most time-consuming of those already described, but advocates claim greater sensitivity and less background staining than with other methods and this seems to be borne out by the experience of most of those who have used it.

Avidin-Biotin Method

Biotin, a water soluble vitamin, is conjugated either to the primary antibody or to the second antibody specific for the species and immunoglobulin class of the primary reagent, either of which can then in turn be detected by avidin linked to a tracer substance (Hsu and Raine 1981). Avidin and biotin have a high and specific affinity for each other. Streptavidin, obtained from *Streptomyces avidinii*, is claimed to demonstrate less non-specific binding.

Hapten Method

Dinitrophenyl (DNP) is covalently linked to a second antibody specific for the species and class of immunoglobulin employed as the primary reagent. The DNP-conjugated second antibody is then located with a pentameric IgM monoclonal antibody specific for DNP, the final step being treatment of the section with DNP-peroxidase conjugate.

Enzyme-Bridge Method

Three antibody layers are employed: the primary antibody is, in the usual way, specific for the antigen being sought; the second antibody is from another species and is specific for the species and immunoglobulin class of the primary antibody; and the third antibody is of the same species as the primary reagent, but is specific for an enzyme such as peroxidase. The final step is exposure of the section to free peroxidase and its demonstration in the usual way with DAB and hydrogen peroxide (Mason et al. 1969). This relatively time-consuming method appears to be little used today.

Labelled-Antigen Method

This method requires the prior preparation of a conjugate comprising the antigen being sought and a tracer (e.g. peroxidase) (Mason and Sammons 1979). The section is treated with one antibody only which, being divalent, binds to the substance in the section and the same substance added subsequently in a conjugated form.

Primary Antibodies

Primary antibodies for immunohistology may be either polyclonal antisera raised in animals or monoclonal antibodies produced by hybridomas in vitro (Kohler and Milstein 1975). Hybridomas generating monoclonal antibodies are produced by fusing lymphocytes from appropriately immunised animals, usually mice, with the cells of a myeloma cell line; the resulting hybrid cells are cloned and then propagated as either ascites tumours or tissue cultures. After fusion and cloning the hybrid cells have the immortality of the parent myeloma cells and the antibody specificity of the immune lymphocytes. These monoclonal antibodies are now available commercially, specific for a wide range of antigens and tending to displace many of the previously used polyclonal antisera.

Because of their superior specificity, monoclonal antibodies tend to give cleaner preparations with less background staining. A common misconception is that most monoclonal antibodies will work only on frozen sections; this is not so. Monoclonal antibodies have enlarged the repertoire of substances detectable by immunohistology to now include those that are especially susceptible to fixation, such as lymphocyte surface markers. The requirement for frozen sections is a property of the antigen rather than the antibody.

Bridging Reagents

Bridging reagents are essential in the tracer-complex methods and desirable when the primary antibody is human (e.g. immune serum after a specific infection). The use of another antibody as the bridging reagent has been described on p. 58 with reference to the tracer-complex method. An alternative is to use staphylococcal protein A, a substance which avidly binds to the Fc portion of immunoglobulin especially of IgG class.

Tracer Substances

Fluorochromes

Fluorochromes were the first tracers to be used in immunohistology. The most commonly used are fluorescein (fluoresces green) and rhodamine (fluoresces red). Fluorochromes absorb light of short wavelength, usually in the ultra-violet part of the spectrum, and emit visible light of longer wavelength. The use of fluorochromes as tracers in immunohistology suffers from important disadvantages: a fluorescence microscope specially equipped with an ultra-violet light source and the appropriate filters and lenses is required; they are difficult to use with paraffin wax sections of fixed tissues because the processing of tissue in this way often results in unacceptable autofluorescence; and the stained preparations are not very permanent because the fluorescence fades and, if unfixed frozen sections have been used, the fluorochrome and antibody may diffuse away from its original location. The relative merits of immunofluorescence and immunohistology with other tracers are summarised in Table 4.1.

Enzymes

It was the conjugation of antibodies to enzymes that permitted the recent advances in immunohistology. The most popular enzyme has been horseradish peroxidase. This catalyses the reaction of 3,3'-diaminobenzidine tetrahydrochloride (DAB) with hydrogen peroxide to give a dark brown reaction product—oxidised DAB. The staining can be optionally intensified by a procedure in which metallic silver is deposited on the final reaction product (Gallyas et al. 1982). DAB has, however, two disadvantages and although these have not as yet caused it to be displaced as the most widely used tracer for immunohistology, they have nevertheless stimulated a search for more acceptable tracers (e.g. a-naphthol pyronin, 2,2'-oxydiethanol 4 chloronaphthol, 3-amino 9 ethyl carbazol). First, DAB is a suspected carcinogen and must be handled with great care to avoid inhalation of the powder when dispensing it or contact with the skin in either its dry or dissolved state. Second, the oxidation of DAB in the presence of hydrogen peroxidase will be catalysed not only by the peroxidase conjugated to the final antibody, but also by endogenous peroxidase within many cells in the body (e.g. erythrocytes, granulocytes). The problem of endogenous peroxidase can be overcome by pretreating the sections appropriately (see p. 62), but it nevertheless adds time and cost to the procedure. An alternative enzyme is alkaline phosphatase, which when reacted with fast red TR as a substrate yields a vivid red-coloured product.

Gold Particles

Gold conjugates are being used increasingly for immunohistology and immuno-electron microscopy (Faulk and Taylor 1971; Geoghegan et al. 1978). The conjugation of antibody molecules to gold particles is a fairly simple matter. The sensitivity of gold immunohistology can be increased by the immunogold-silver technique in which the gold particles act as a seed for the deposition of silver in a process akin to photographic development (Holgate et al. 1983).

Double-Labelling Techniques

Sometimes it is desirable to locate two different substances in the same section, to compare their distribution for example. This can be done by using two different tracers (e.g. DAB and alkaline phosphatase) in two consecutive immunostaining procedures on the same section.

Fixation

Routine methods of fixation employed in most surgical pathology laboratories are satisfactory for most immunohistological investigations by non-fluorescent methods. (The problem of autofluorescence induced by fixation has been referred to already). This should not, however, deter one from evaluating a range of fixatives to determine the optimum method for a specific study.

The problem with most fixatives is that they modify the tissue substances in such a way as to obscure their antigenicity and thus render them unrecognisable to the respective antibodies. This effect can be reversed by limited proteolysis with an enzyme such as trypsin. Limited proteolysis reverses the cross-linking of proteins induced by fixatives. The optimum concentration and/or time must be determined first by titration, preferably guided by the effect on some antigen that is ubiquitous, such as factor-VIII-related antigen in vascular endothelium (Fig. 4.2).

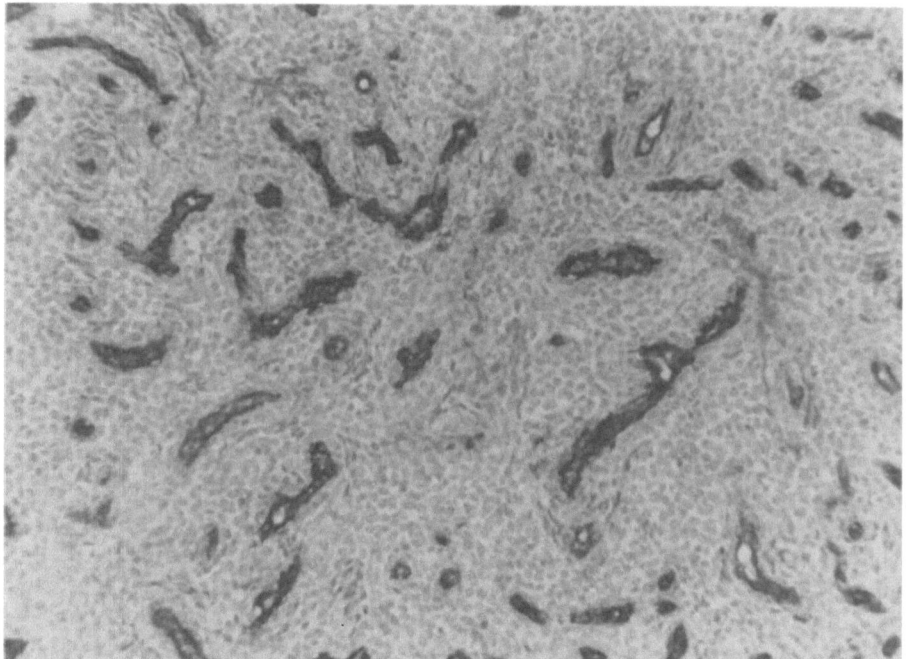

Fig. 4.2. Haemangiopericytoma stained for factor-VIII-related antigen to demonstrate the endothelium of the abundant stromal vessels. Factor-VIII-related antigen is an almost ubiquitous marker of vascular endothelium. Peroxidase-antiperoxidase, × 148

Use of Enzymes to Unmask Antigens

Cross-linking of proteins, induced by fixatives such as formalin, can cause the masking of antigenic determinants normally recognisable by antibodies applied to sections of unfixed tissue. To some extent the diminished intensity of staining that results can be overcome by using inordinately high concentrations of the primary antibody, but this is usually accompanied by an unacceptable degree of background staining. This problem can be largely eliminated by pretreating the tissue sections before immunostaining with a proteolytic enzyme; trypsin (Huang et al. 1976; Curran and Gregory 1977), protease type VII (Denk et al. 1977) and pepsin (Dixon et al. 1980) have been used successfully. The incubation conditions—time, temperature, enzyme concentration—must be very carefully controlled; undertreatment will not have the desired effect and overtreatment may result in digestion of the antigenic site being sought or even disintegration of the entire section.

Endogenous Peroxidase

A particular problem with the use of peroxidase as a tracer is that peroxidase itself is present also in many cells and tissues. This can lead to confusion because peroxidase staining may be due to either the peroxidase tracer used in the immuno-staining procedure or the presence of endogenous peroxidase in the tissue section or cells. Peroxidase is abundant in erythrocytes, granulocytes and hepatocytes, including many other cells. Although the procedures used to block endogenous peroxidase carry some risk of disturbing the antigenicity of tissue substances, it has become standard practice except when working with frozen sections. The methods available include hydrogen peroxide in methanol or periodic acid/sodium borohydride (Heyderman and Neville 1977). Failure to block endogenous peroxidase may give the appearance of false positive immunohistological staining.

Controls

Ideally, the specificity of all primary antibodies or antisera should be fully validated before use, especially if clinical decisions are to be based on the results. This is best tested by absorption with the antigen for which specificity is claimed; specific immunostaining should be abolished. Most users of commercial reagents will usually accept the supplier's accreditation of the specificity and rely on internal controls in the tissue sections. For example, if an antibody claimed to be specific for cytokeratin stains lymphocytes or vascular endothelium in a tissue section it obviously either does not have the claimed specificity or nonspecific staining is being produced by some flaw in the method.

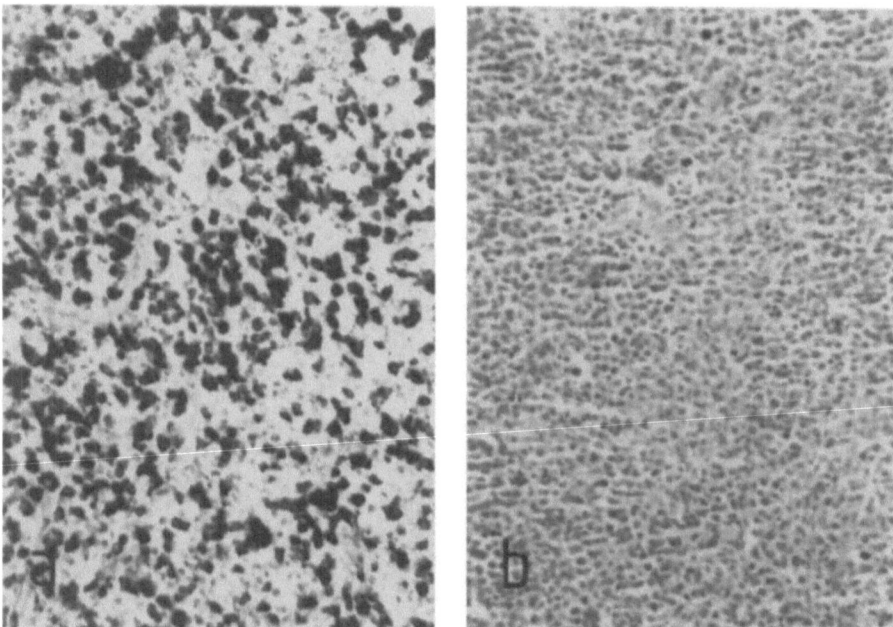

Fig. 4.3a,b. Light-chain restriction confirms B-cell lymphoma. a Abundant cytoplasmic staining with anti-serum to kappa light chain. b Antiserum to lambda light chain produces virtually no staining, thus showing the monoclonal (i.e. neoplastic rather than reactive) nature of the lesion. Peroxidase-antiperoxidase method on paraffin sections of formalin-fixed tissue, × 160

Common Applications

Investigation of Lymphoproliferative Disorders

Immunohistology has had its greatest impact on the diagnosis and classification of lymphomas. The first question that often has to be asked is "is it lymphoma?", and this can be answered least equivocally by immunohistology if the process is of B-lymphocyte lineage. Neoplastic B-cell proliferations show light chain restriction; that is, because the cells are derived from a single clone, they all synthesise the same type of immunoglobulin light chain—either kappa or lambda, never both. Reactive B-cell lesions are polytypic and will be populated by a mixture of cells each containing either kappa or lambda light chains. This can be done successfully in most cases on paraffin wax sections of routinely fixed and processed tissue (Fig. 4.3), but this should not excuse one from making every effort to secure fresh tissue.

For proper evaluation of a lymphoproliferative problem it is now considered essential that the tissue is delivered to the laboratory fresh and as soon as possible after removal. This enables frozen sections of the unfixed tissue to be cut for staining with monoclonal antibodies to lymphocyte surface markers (Fig. 4.4).

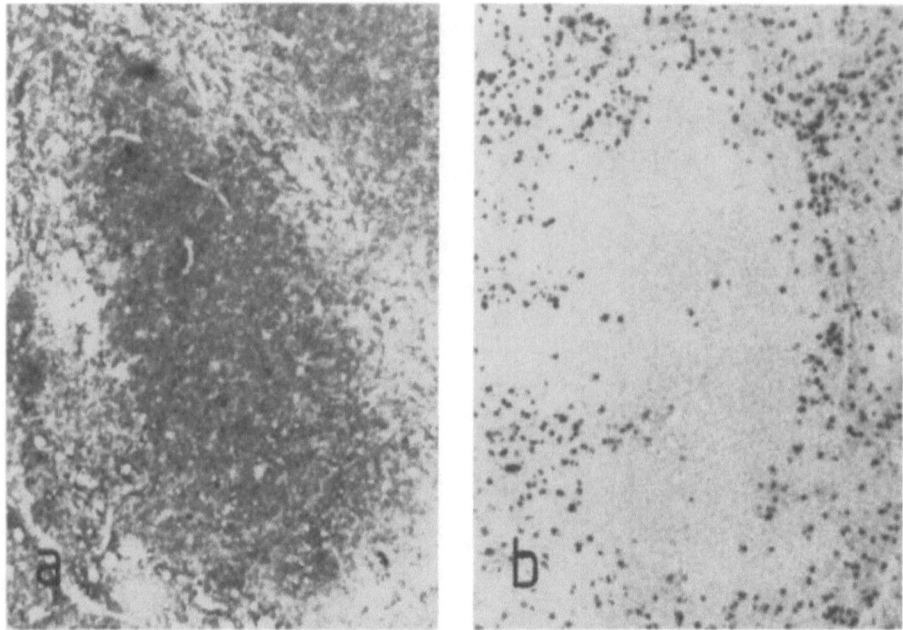

Fig. 4.4a,b. Confirmation of a follicular B-cell (centrocytic) lymphoma by monoclonal antibodies applied to cryostat sections. **a** Monoclonal antibody to a B-lymphocyte surface marker produces uniform staining of the follicular nodules of neoplastic centrocytes. **b** Reciprocal pattern of staining produced by a monoclonal antibody to a subpopulation of T-lymphocytes in the interfollicular tissue. Indirect immunoperoxidase method, × 99

Determination of Tumour Histogenesis

Immmunohistology is beginning to displace electron microscopy as the next step after routine light microscopy for the histogenetic diagnosis of tumours (Gatter et al. 1985). There are as yet, however, no useful "cancer antigens" that might enable one to diagnose malignancy in doubtful cases; recent claims for cancer-specific antibodies for use in immunohistology have not stood up to critical appraisal. So the diagnosis of malignancy must be made in some other way, usually by the application of standard histopathological criteria. The only situation in which immunohistology can alone lead to a diagnosis of malignancy is by the demonstration of antigen-bearing cells foreign to the tissue being examined. For example, if cytokeratin is demonstrated by immunohistology in a group of cells in a lymph node this would strongly suggest a metastasis since there are no cells within a normal lymph node that contain cytokeratin.

A fairly common diagnostic problem is the distinction between "anaplastic" carcinoma and lymphoma; both can produce a rather nondescript histological picture lacking immediately obvious differentiated characteristics. But one of the easiest ways of distinguishing these two entities is to immunostain one section for cytokeratin and another for leucocyte common antigen; these will stain respectively carcinoma and lymphoma in most instances (Fig. 4.5). These and other commonly used markers are listed in Table 4.2.

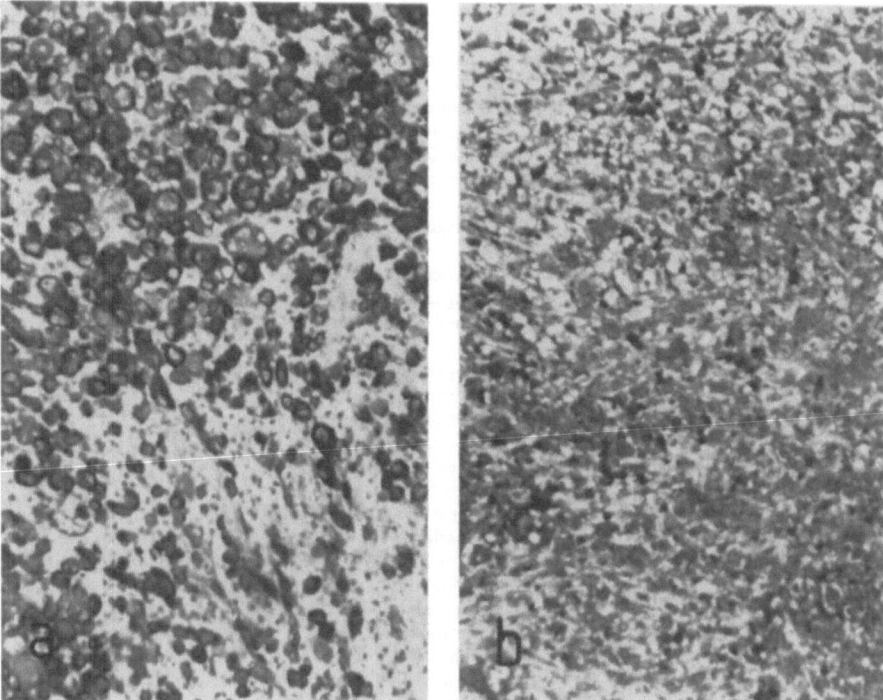

Fig. 4.5a,b. Poorly differentiated carcinoma distinguished from lymphoma by the use of monoclonal anti-bodies to cytokeratin and leucocyte common antigen (LCA). a Cytokeratin antibody produces cytoplasmic staining of the large pleomorphic neoplastic cells. b LCA antibody stains only occasional infiltrating stromal inflammatory cells. Peroxidase-antiperoxidase method, × 150

Renal Biopsies

Immunohistology has become mandatory for the proper interpretation of renal bi-opsies from patients with glomerular disease. In many institutions this is still per-formed by immunofluorescence, but the permanence of immunohistochemical preparations offers distinct advantages, particularly if the biopsy needs to be reviewed and discussed at a later date, as is often the case (Fig. 4.6). The application of immunoperoxidase techniques to renal biopsies has been reviewed by MacIver and Mepham (1982).

Problems of Interpretation

It is easy for the novice to assume that the result of an immunohistological investiga-tion is either "positive" (i.e. "brown" if DAB is the tracer) or "negative" (i.e. coun-terstain only), but all users of these techniques, however experienced, need to be wary of jumping to hasty conclusions after just a cursory glance at the section. False-positive staining may result from failure to block endogenous peroxidase, in which case there

Table 4.2. Some markers commonly used for the immunohistological diagnosis of tumours

Marker specificity of antibody or antiserum	Diagnostic implications of marker
a_1-Antitrypsin	Histiocytic lineage (granular perinuclear staining) and some T-lymphocytes
a-Fetoprotein	Endodermal sinus differentiation
	Liver-cell carcinoma
Carcinoembryonic antigen	Many adenocarcinomas; mesothelial cells usually negative
Cluster differentiation (CD) antigens:	
CD1 (e.g. OKT6)	Cortical thymocytes
	Langerhans' cells
CD3 (e.g. OKT3)	Pan-T-lymphocytes
CD4 (e.g. OKT4, Leu3)	Helper-T-lymphocytes
CD8 (e.g. OKT8, Leu 2a)	Suppressor/cytotoxic T-lymphocytes
CD20 (e.g. Bl)	B-lymphocytes
CD11 (e.g. Mo1)	Myeloid lineage
	Monocyte/macrophage lineage
CD15 (e.g. Leu M1)	Myeloid lineage
	Reed-Sternberg cells
Cytokeratin (CAM 5.2, etc.)	Epithelium or mesothelium
Desmin	Muscle
Epithelial membrane antigen	Epithelium or mesothelium (some lymphomas rarely)
Factor VIII-related antigen	Endothelium
γ-Enolase (NSE)	Neuroendocrine cells
	Melanocytes
Glial fibrillary acidic protein	Glial cells
Hormones (e.g. thyroglobulin, HCG, calcitonin)	Identifies cell as probably synthesising demonstrated hormone
Ig heavy chains ($a, \delta, \varepsilon, \gamma, \mu,$)	Immunoglobulin-producing cell
	Typing of myeloma
Ig light chains (κ, λ)	Polytypic (reactive) or monotypic (neoplastic) lymphoid cells
Leucocyte common antigen	White cell series
Lysozyme	Histiocytic lineage (granular perinuclear staining)
Myoglobin	Muscle
Prostate-specific antigen	Prostatic epithelium
Prostatic acid phosphatase	Prostatic epithelium
	Some hindgut carcinoids
S100	Nerves, chondrocytes, melanocytes, Langerhans' cells, APUD cells
Vimentin	Most connective tissues

will be apparent staining of the ubiquitous erythrocytes and granulocytes. Nonspecific staining may result from passive uptake of a circulating serum protein by dead or dying cells. False-positive staining for immunoglobulin may therefore be seen in the cytoplasm of degenerate cells that do not normally synthesise or store it. This staining due to passive uptake usually results in diffuse cytoplasmic staining rather than the more specific granular staining seen in the Golgi region of cells that have synthesised the material.

Internal controls are present in most tissues. For example, if one is seeking cytokeratin in a skin tumour, there should be positive staining in the adjacent epidermis as confirmation that the staining procedure has been performed correctly. Indeed, when selecting blocks for immunohistology it is helpful to bear in mind the usefulness of having internal control tissues in the final preparation.

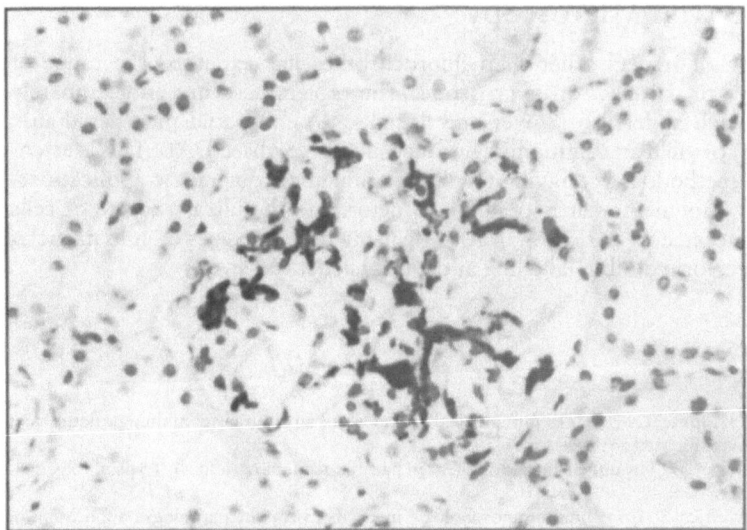

Fig. 4.6. Immunohistology of a renal biopsy showing predominantly mesangial deposition of IgA. Peroxidase-antiperoxidase, × 345

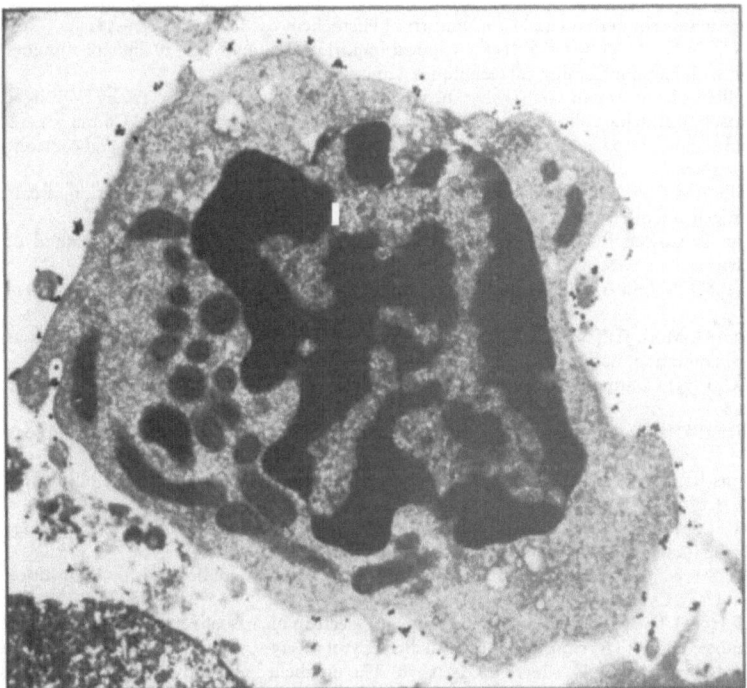

Fig. 4.7. Immunoelectron microscopy using gold-conjugated antibody. Numerous electron-dense gold particles are now associated with the surface membrane marker on this lymphoid cell. Indirect immunogold method, × 1606

Immuno-electron Microscopy

The introduction of tracers other than fluorochromes has extended the range of immunohistology to the electron microscope. The tracers used are either innately electron-dense, such as ferritin (Singer and Schick 1961) and gold particles (Faulk and Taylor 1971) or yield an osmiophilic product such as oxidised DAB. This is essentially a research method, as yet having few if any immediate diagnostic applications. Nevertheless, by this method it is possible to determine the ultrastructure of cells of known immunophenotype (Fig. 4.7). Also by using different sizes of gold particles it is possible to perform double-labelling at the ultrastructural level.

References

Coons AH, Creech HJ, Jones RN (1941) Immunological properties of an antibody containing a flourescent group. Proc Soc Exp Biol Med 47: 200–202

Curran RC, Gregory J (1977) The unmasking of antigens in paraffin sections of tissue by trypsin. Experientia 33: 1400–1401

Denk H, Seyre G, Weirich E (1977) Immunomorphologic methods in routine pathology: application of immunofluorescence and the unlabelled peroxidase-antiperoxidase technique to formalin fixed, paraffin embedded kidney biopsies. Beitr Pathol 160: 187–197

Dixon AJ, Burns J, Dunnill MS, McGee J (1980) Distribution of fibronectin in normal and diseased human kidney. J Clin Pathol 33: 1021–1028

Faulk WP, Taylor GM (1971) An immunocolloid method for the electron microscope. Immunochemistry 8:1081–1083

Gallyas F, Gorcs T, Merchenthaler I (1982) High-grade intensification of the end-product of the diaminobenzidine reaction for peroxidase histochemistry. J Histochem Cytochem 30: 183–184

Gatter KC, Alcock C, Heryet A, Mason DY (1985) Clinical importance of analysing malignant tumours of uncertain origin with immunohistological techniques. Lancet I: 1302–1305

Geoghegan WD, Scillian JJ, Ackerman GA (1978) The detection of human B-lymphocytes by both light and electron microscopy utilizing colloidal gold-labelled anti-immunoglobulin. Immunol Comm 7:1–12

Heyderman E (1979) Immunoperoxidase technique in histopathology: applications, methods, and controls. J Clin Pathol 32: 971–978

Heyderman E, Neville AM (1977) A shorter immunoperoxidase technique for the demonstration of carcinoembryonic antigen and other cell products. J Clin Pathol 30: 138–140

Holgate CS, Jackson P, Cowen PN, Bird CC (1983) Immunogold-silver staining: a new method of immunostaining with enhanced sensitivity. J Histochem Cytochem 31: 938–944

Hsu S–M, Raine L (1981) Protein A, avidin and biotin in immunohistochemistry. J Histochem Cytochem 29: 1349–1353

Huang SN, Minassian H, More JD (1976) Application of immunofluorescent staining on paraffin sections improved by trypsin digestion. Lab Invest 35:383–390

Kohler G, Milstein C (1975) Continuous culture of fused cells secreting antibody of predefined specificity. Nature 256:465–467

MacIver AG, Mepham BL (1982) Immunoperoxidase techniques in human renal biopsy. Histopathology 5:249–267

Mason DY, Sammons RE (1979) Labelled antigen method of immunoenzymatic staining. J Histochem Cytochem 27:832–840

Mason TE, Phifer RF, Spicer SS, Swallow RA, Dreskin RB (1969) An immunoglobulin enzyme-bridge method for localizing tissue antigens. J Histochem Cytochem 17: 563–569

Nakane PK, Pierce GB (1966) Enzyme-labelled antibodies: preparation and application for localization of antigens. J Histochem Cytochem 14: 929–931

Singer SJ, Schick AF (1961) The properties of specific stains for electron microscopy prepared by conjugation of antibody molecules with ferritin. J Biophys Biochem Cytol 9: 519–537

Sternberger LA, Hardy PH, Cuculis JJ, Meyer HGH (1970) The unlabelled antibody–enzyme method of immunohistochemistry: preparation and properties of soluble antigen-antibody complex. J Histochem Cytochem 20: 315–333

Taylor CR (1978) Immunoperoxidase techniques: practical and theoretical aspects. Arch Pathol Lab Med 102: 113–121

5 Interpretation of Histological Appearances

The human brain has a remarkable capacity to compute a diagnosis from visual information. It matters not for these purposes whether the individual is diagnosing the identity of a friend met in the street, a brand of food on the supermarket shelf, or a histological section of a colorectal adenocarcinoma; the diagnostic process is the same in each instance; what distinguishes each example is the previous experience of the observer that enables diagnostic recognition. A remarkable aspect of diagnostic recognition is the enormous amount of information present within visual images that the human eye can perceive and the human brain can compute. For example, it has been estimated that a single microscope field contains at least 500 000 picture points resolvable by the human eye and that 16 levels of grey-value can be discriminated for each picture point and for the colours red, green, and blue and intermediate hues (Bartels and Wied 1981). Beyond that relatively simple photometric information there is the higher order information within the image—size, shape, orientation, textures, relative positions, frequency of components (Julesz et al. 1973; Julesz 1975). Then there is the comparison to be made with previously experienced images and interpolation with the patient's clinical history, biochemistry, immunology, etc. as relevant.

Sometimes the nomenclature we use betrays the unity between the everyday recognition skills used by the general public and the apparently more sophisticated recognition skills of the trained histopathologist. This is exemplified by the names we give to certain histological details; the names are often derived from a resemblance to more familiar objects in public life: e.g. "coffee-bean nuclei" in Brenner and granulosa cell tumours of ovary; "oat-cell carcinoma" as a synonym for small-cell carcinoma of lung; "Orphan Annie nuclei" in papillary carcinoma of the thyroid (named after the forlorn and empty eyes of a once popular cartoon character in North American newspapers).

Artifact of Sections

Although gross morphology remains very important, the histological section has evolved as the main diagnostic medium of the histopathologist. Sections, cut so thin that they can be stained and permit the transmission of light, are essentially two-dimensional samples of three-dimensional tissues. A section is therefore in a sense an artifact, but an artifact so consistent as to be reliably useful. This two-dimensional artifact must be translated into three dimensions in the mind of the observer in order to provide a full understanding of the true structure of the lesions present. The consequences of literally interpreting two-dimensional images are well illustrated by the somewhat facile temptation to describe the downward projections of the epidermis as pegs rather than ridges. Three-dimensional preparations, in which the epidermal-dermal interface can be seen, clearly demonstrates ridges, not pegs, though they appear to be peg-like in thin sections (Pinkus 1960). It is also naive to describe a section of a spindle-cell tumour as composed of a mixture of fusiform and round cells, unless there really are two discrete cell populations; fusiform cells appear round in transverse section. Similarly, "polyhedral" is a more accurate term to use than "polygonal" when describing the shape of the cells of an otherwise indescribable anaplastic tumour.

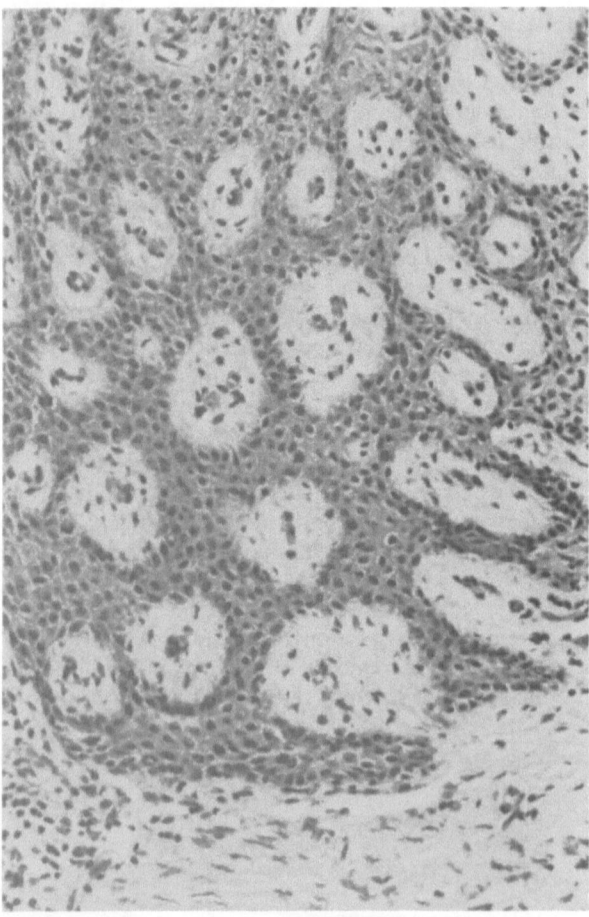

Fig. 5.1. Tangential section of skin showing that the peg-like downward projections of the normal epidermis seen in perpendicular sections are, in reality, interconnecting ridges. In a small poorly orientated biopsy this appearance might be confused with neoplastic invasion. Haematoxylin and eosin, × 215

After some experience the microscopist adapts to the two-dimensional world of tissue sections and can usually subconsciously compute the third dimension from the available information. However, the mind is occasionally perplexed if the orientation of the plane of section is unfamiliar. Tangential sections of skin, for example, can give the false impression that irregular strands of squamous epithelium are "invading" dermal connective tissue (Fig. 5.1). With skin and mucosal biopsies, the tissues should always be embedded so that the plane of section will be perpendicular to the surface.

Basic Microscopy

"Beware of determining and declaring your opinion suddenly on any object; for imagination often gets the start of judgement, and makes people believe they see things, which better observations will convince them could not possibly be seen: therefore assert nothing till after repeated experiments and examinations in all lights and in all positions.

When you employ the microscope, shake off all prejudice, nor harbour any favourite opinions; for, if you do, 'tis not unlikely fancy will betray you into error, and make you think you see what you would wish to see.

Remember that truth alone is the matter you are in search after; and if you have been mistaken, let not vanity seduce you to persist in your mistake".

(from "The Microscope Made Easy" by Henry Baker 1742).

Knowledge of the principles of the light microscope is assumed (James 1976). This section is devoted to a brief consideration of the use of the microscope as a diagnostic tool. The microscope should be properly adjusted and have clean lenses. The environment is equally important; background noise and repeated interruptions are not conducive to diagnostic work.

The optics of the microscope can be modified to extend the range of usefulness of the instrument. Though dark-ground or phase-contrast illumination are rarely used in diagnostic work on sections of fixed tissues, polarising filters are often used to detect birefringent particles or crystals (Wolman 1970) and the specificity of Sirius red or Congo red stains for amyloid can be increased by demonstrating a red-green colour shift (dichroism) in a rotating polarised beam.

Every available section should be examined initially by scanning with low- or medium-power lenses. With mechanical stages it is easy to scan across the section in a stepwise fashion. A hand lens, or even the low power of a dissecting microscope, is a useful way of assessing the overall tissue architecture. Trainees should avoid the temptation to spend a disproportionate amount of time on the high-power appearances. Most disease processes can be recognised on low- or medium-power microscopy; the higher ranges of the microscope (e.g. × 25, × 40 objectives) often merely provide confirmatory evidence. On the other hand, the oil-immersion objective (× 100) is probably underused by many pathologists, especially those who reserve it only for detecting micro-organisms. It is, for example, useful for observing nuclear clefts and convolutions in lymphomas and for seeking the intercellular prickles that often denote squamous histogenesis in otherwise poorly differentiated tumours.

So that important features are not overlooked, the evaluation of a tissue by histology should be done in a systematic sequence. For example, in renal biopsies— glomeruli, tubules, vessels, interstitium; in skin biopsies—epidermal strata, papillary

dermis, reticular dermis, subcutis, vessels, adnexae; in mucosal biopsies—gland architecture, crypts, lamina propria, submucosa; in liver biopsies—portal tracts, lobules,
architecture, sinusoids; in a tumour—histogenesis, local extent, stroma, vascular permeation; etc.

Before specialising in histopathology the trainee will have had some basic instruc-
tion in pathology during the undergraduate medical course and should be capable
of writing a purely descriptive report of what is seen in a gross specimen or tissue
section. Senior colleagues can be asked to comment on the written description and
to point out any omissions. This is a helpful way of ensuring that one has perceived
all the important features of a specimen. Interpretation is a separate exercise which
cannot be done accurately unless all the salient features in a biopsy are first recognised.

The actual interpretative process should be logically conducted, whether the diagnosis is made by heuristic analysis or by pattern recognition (see pp. 75–79). For
example:

1. Is this appearance normal or abnormal?
2. If abnormal, what disease process is present?
3. Could anything else explain these appearances?
4. Is there any co-existent lesion?

If confronted with a difficult biopsy it may be helpful to identify the nature of
the difficulty. Different types of problem require different methods for their solution.

1. Is the difficulty due to poor fixation, autolysis, faulty processing, or technical
   artifact?
2. Is the clinical information adequate?
3. Could it be just a small biopsy of a common lesion?
4. Could it be a very rare lesion, or perhaps a common lesion that is only rarely
   biopsied?
5. Has the lesion only partly evolved?
6. Could the disease have been modified by treatment or by some other co-existent
   disease?
7. Is the lesion of a borderline nature (e.g. normal versus marginal abnormality, or
   grades of abnormality such as dysplasias)?
8. Does the histological picture of one disease mimic that of another?
9. Are there some very conspicuous morphological features which, though diagnostically unimportant, are distracting from the real problem?

Artifacts in Sections

Often trainees and sometimes even experienced histopathologists encounter objects
or other features in tissue sections which can be baffling or misleading—and yet they
are just artifacts. Some are inconsequential; others mimic, confuse or obscure the

morphology of the disease. They have been comprehensively reviewed by Thompson and Luna (1978) and by Wallington (1979).

Some of the more troublesome artifacts commonly seen in routine histopathological practice include the following:

Crushing. Cells and tissues are extremely fragile when fresh and must be handled with great care by both the clinician who has taken the biopsy and the pathologist who receives and examines it in the laboratory if a confusing result is to be avoided. For example, lymphocytes and the cells of oat-cell carcinoma of the bronchus are very susceptible to crushing and, in small bronchial biopsies, crushed lymphocyte aggregates can be easily mistaken for crushed oat-cell carcinoma and vice-versa.

Cautery and Diathermy. Diathermy is often used to assist in the removal of tissue, particularly where bleeding is a problem, and in some situations, such as the prostate, it is the principal method by which the tissue is resected endoscopically. The margins of pieces of tissue removed in this way will inevitably suffer damage from the heat of the diathermy process and this will often obscure tissue detail. Sometimes the nuclei in the diathermied areas appear abnormally hyperchromatic and, if the tissue architecture is distorted, this may be misinterpreted as evidence of neoplasia.

"Floaters". Extraneous pieces of tissue are a nuisance, especially when they contaminate small biopsies. These unwelcome intruders, often referred to as "floaters", may contaminate biopsies at the following stages in tissue handling and processing: when the biopsy is first handled in the laboratory it may become contaminated with fragments of other tissue(s) previously in contact with the instruments or surfaces on which the biopsy is placed; when the biopsy is being processed into paraffin wax it may become contaminated with small fragments of other tissues being processed in the same batch; and when the tissue sections are being floated out on a water bath they may pick up residual fragments of other sections previously floated out on that bath. The best way to avoid these "floaters" is to pay meticulous attention to cleanliness at all stages of specimen handling and processing. For example, when handling and describing a batch of specimens it is best to start with the smallest and least friable biopsies and gradually work up in size to those that are most likely to shed fragments of potentially contaminating tissue, such as tumour resections. "Floaters" in tissue sections are easily recognised as extraneous when the association is implausible (e.g. a fragment of a fibroadenoma of the breast contaminating a rectal biopsy), but they can give rise to considerable difficulties if the association is plausible (e.g. a fragment of adenocarcinoma from a colectomy specimen contaminating a rectal biopsy from another patient); misreporting the latter rectal biopsy as "adenocarcinoma" may have dire consequences if none is actually present.

Foreign Bodies. Foreign material may be picked up while the tissue is being handled prior to biopsy or while it is still fresh; starch from glove powder entrapped in fibrin, for example, is commonly seen on the surface of many biopsies. Wood fragments from the surface on which the tissue is handled and dissected may also be seen in histological sections. It is fairly easy to identify these particles of foreign material as artifactual because there will be no associated tissue reaction (e.g. foreign body giant cells). Nevertheless, they are best avoided. Pollen grains frequently contaminate the sticky mountant used when the stained section is finally protected by placing a coverslip on it, especially in the summer months; these may be misinterpreted as pathogenic fungi, but critical examination will reveal that the grains lie outside the plane of the section.

"Chatter" and Scoring. These are annoying technical problems that invite technical solutions. Scoring can be minimised by careful examination of the tissue before it is processed to make sure that no calcified material, clips or sutures are included. Lightly calcified tissue can be processed in the usual way, but anything more than that needs decalcification.

Pigments. Artifactual pigments are usually recognisable with experience. The most common is formalin pigment, an artifact of formalin fixation particularly of very vascular organs such as the spleen. It is also more common in tissues which have been in formalin for a long time and is therefore most commonly seen in post-mortem histology. It can be removed with alcoholic picric acid, but this is rarely necessary unless one wants a near perfect section for teaching, photography, etc. Some lipids in tissues will also react with formalin to yield a brown finely-granular pigment. Pigments encountered in histological sections are summarised in Table 5.1.

Table 5.1. Some pigments encountered in histological sections

Pigment	Colour	Stain	Significance
Bile	Yellow	Gmelin	E.g. cholestasis
Carbon	Black		Anthracosis
Ceroid	Yellow	PAS	E.g. in Kupffer cells, recent liver injury
Formalin[a]	Black		Artifact of formalin fixation
Haematoidin	Bright yellow		Organised haematoma
Haemosiderin	Light brown	Perls'	Organised haematoma, haemochromatosis, etc.
Lipofuscin	Dark brown	Ziehl-Neelsen Schmorl reaction	"Brown atrophy", melanosis coli, etc.
Melanin	Dark brown	Masson-Fontana	Melanoma, etc.
Malaria[a]	Black		Malaria, *S. mansoni*
Polymerised homogentisic acid	Black		Ochronosis
Silver	Black		Argyria[b]

[a] Removed by alcoholic picric acid.
[b] Usually deposited on basement membranes.

Methods of Interpretation

It would be easy and quite misleading to avoid the question of interpretative methods by simply saying, for instance, that a lymph node is reactive because it *looks* reactive. Certainly after sufficient exposure to lymph node histology, the trainee will usually perceive various features common to all lymph nodes that *look* reactive to more experienced pathologists. However, it is probably unwise, even for the most accomplished pathologist, to be utterly convinced of the diagnosis simply because a lesion *looks* like what it is. Nevertheless, pattern recognition is one of the most widely used methods of histopathological diagnosis and has much to commend it in terms of economy of effort and accuracy, provided that the user is aware of the actual psychological processes involved and the dangers of being deceived.

Diagnostic processes can be categorised as either "pay off", heuristic, or pattern recognition (Dudley 1968). The "pay off" method has no direct application to histopathology. It is essentially a therapeutic solution to a diagnostic problem. If the patient's right iliac fossa pain and tenderness responds to appendicectomy, then the diagnosis is probably appendicitis rather than mesenteric adenitis or salpingitis. It would be unwise to test the validity of a biopsy diagnosis by seeing what happened when the physician or surgeon acted upon the report, though in the long term it is valuable to review biopsies in the light of responses to therapy. This may even result in a reconsideration of that type of biopsy problem, and perhaps define new criteria that could help to predict the response to various treatments.

The heuristic method and pattern recognition are, however, the main diagnostic processes used in biopsy interpretation.

Heuristic Analysis

The logic of discovery, invention, and problem solving was given the name "heuretic" by Sir William Hamilton (1788–1856), a Scottish philosopher. The derived term "heuristic" describes goal-seeking behaviour. An obvious prerequisite for heuristic analysis is the definition of the actual goal, target or problem that needs solving. "Does the patient have carcinoma of the rectum?" "Is there any parakeratosis?" "Is this liver cirrhotic?" "Is this breast lump a fibroadenoma?" Each question clearly defines a diagnostic problem and, except in cases of appreciable doubt, it would be possible to answer either affirmatively or negatively.

A major disadvantage of the heuristic approach is that posing a specific but misleading question may temporarily lead to neglect of the real problem. If we follow the heuristic method we then have to pose a second specific question, and so on. At the risk of making this exercise simply ridiculous, if the breast lump is *not* a fibroadenoma then what is it? The pattern recognition method applied by the experienced pathologist may arrive at the right diagnosis much more rapidly than if heuristic logic had been used.

What then is the role of heuristic analysis in biopsy interpretation? Trainees and their tutors find this method easier to teach and to learn from than pattern recognition. First, this methodical and objective approach of accumulating information about a biopsy, until the diagnosis is reached, is easier to grasp than the less tangible and more subjective process of pattern recognition. Eventually trainees will have had enough exposure to a variety of lesions to enable them to use the pattern recognition method with increasing frequency. Secondly, all histopathologists ultimately resort to the heuristic method when faced with an apparently insoluble diagnostic problem. A diagnostic flow sheet may be mentally devised to pose a sequence of heuristic challenges. The correct diagnosis may be one that is reached by the shortest path in the heuristic maze, or by a longer path if the result is more consistent with the clinical picture (Fig. 5.2). Lastly, sometimes a diagnosis is made intuitively, but even this should be tested for validity by heuristic logic.

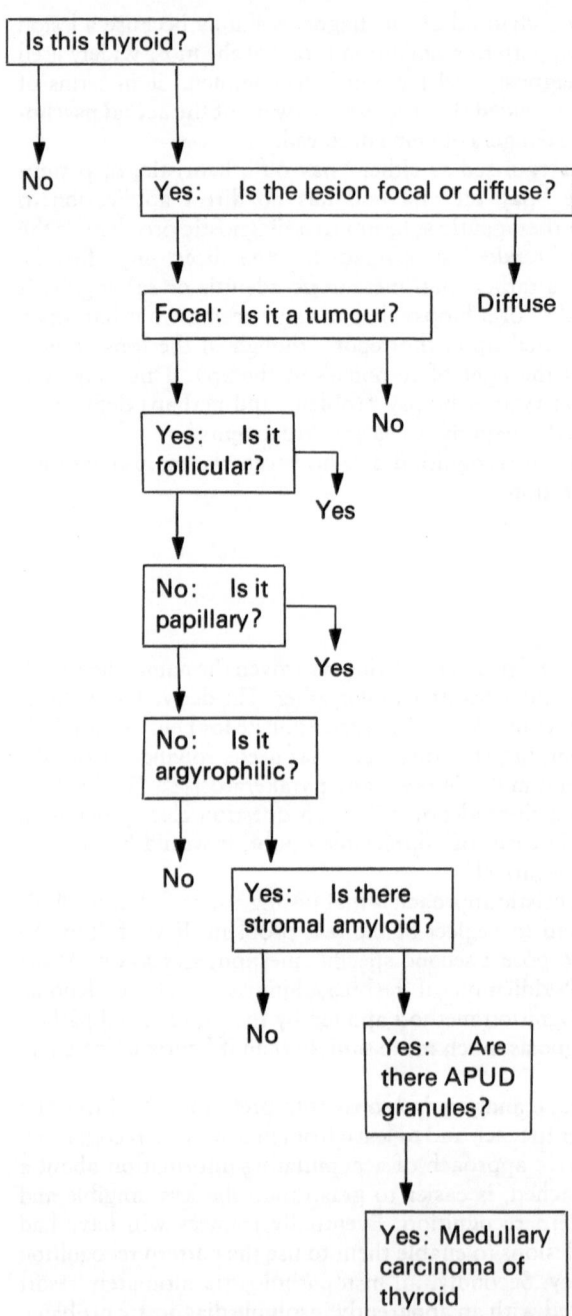

Fig. 5.2. Flow diagram illustrating the diagnosis of a thyroid lesion by heuristic analysis. Pattern recognition may be involved in some of the steps (e.g. recognition of the presence or absence of a follicular pattern), but the overall concept is one of logical stepwise progression. Further confirmation of the lesion as a medullary carcinoma might be sought for by demonstrating calcitonin in the tumour cells by immunohistochemistry.

Pattern Recognition

Although pattern recognition is the most frequently used diagnostic process it does not readily lend itself to systematic explanation. Whereas the stepwise analysis of a histological problem is governed by heuristic logic, pattern recognition involves Gestalt psychology.

Gestalt psychology, based on concepts formulated in Germany at the end of the nineteenth century, is the psychology of the perception of form, as opposed to the identification of an object from the sum of its separate components. In the context of biopsy interpretation this means recognising a lesion from its overall pattern rather than from the sequential assessment of individual features (e.g. mitotic activity, pleomorphism, necrosis, pigment, inflammatory infiltrate). Pattern recognition is more rapid than heuristic analysis, certainly much more subjective, and probably more prone to error. It can be used only for lesions that are sufficiently common or constant in their appearance to be imprinted on the memory. If rare, the memorised pattern may be from a book, atlas or paper describing the lesion rather than from first-hand experience.

Pattern recognition in biopsy interpretation is not a new personal skill that has to be acquired but rather the application of an established skill to an entirely new medium. There are analogies with many activities, from art (Dudley 1968) to ornithology (Ackerman 1978). The music of Beethoven can be distinguished from that of Bartok by the general form of the sound; the listener must, of course, be already acquainted with each composer's style. Similarly, paintings by Renoir can be distinguished from those by Cezanne if one is familiar with the general style of each artist's works. Although Renoir painted everything from nubile nudes to lush landscapes, his work is generally characterised by a certain style that is easily recognised. So it is that the experienced eye can readily and rapidly identify an adenocarcinoma from its overall pattern, even though all adenocarcinomas arising in the same organ differ in detail (Fig. 5.3). The artist's style is analogous to the disease; the artist's subject is analogous to the tissue. The psychology of Gestalt applies equally to the recognition of a piece of music as by Beethoven, of a bird as a sparrow, of a painting as by Renoir, and of a rectal tumour as an adenocarcinoma. This concept is unavoidably pretentious, but the intention is to show that pattern recognition is not a process peculiar to biopsy interpretation.

The experienced histopathologist responds by a subconscious reflex action, recognising a pattern in a biopsy by comparing the lesion in question with a set of memorised patterns accumulated from previous experience. This, "the beholder's share" (Gombrich 1972), enables a diagnosis to be inferred. In contrast, the heuristic approach more often involves reference to a set of external criteria—books, atlases, monographs or perhaps slides in a reference collection. (A reference slide collection, injudiciously used, has the inherent danger that the previous mistaken interpretations may be made with increasing frequency and confidence!) Each individual pathologist has a threshold which sets the degree of similarity that must exist between the morphological feature in question and the reference criteria.

Unquestioning reliance on pattern recognition can have its disadvantages. First, the mind is easily biased by the first impression of a biopsy; this is more likely to be gained through relatively rapid pattern recognition that the more laborious heuristic method. Many biopsy interpretative problems can be solved only through a detailed critical analysis of the appearances, and it may be difficult eventually to accept the correct diagnosis if the mind has been initially seduced by the first brief glimpse.

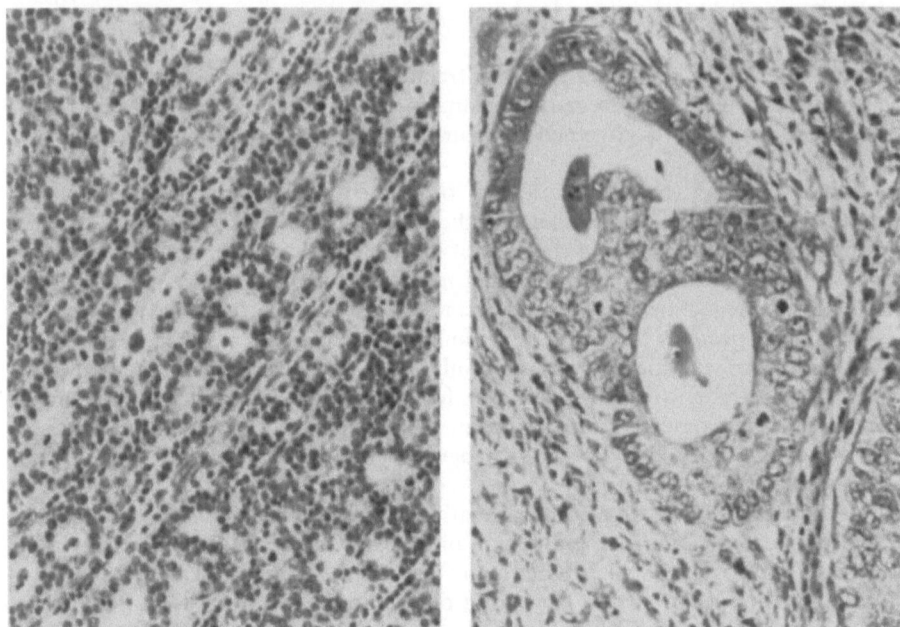

Fig. 5.3. Two gastric tumours differing in several respects (e.g. cellularity, nucleo-cytoplasmic ratio, degree of gland formation, stromal reaction). Both are identifiable, by the experienced observer using pattern recognition, as adenocarcinomas. Less differentiated tumours would necessitate a heuristic approach. × 215

Secondly, just as birds, butterflies or faces are "seen" in the random images of Rorschach ink-blots used for psychological testing, so the prepared mind may read an incorrect but anticipated diagnosis into the element of randomness existing in the histological pattern of another disease. Few lesions are so absolutely typical that they conform exactly to some idealised description. Diversity of histological features is inevitable. The diversity is largely random in nature and may be sufficiently great for the histological pattern of one disease to simulate that of another. Lastly, without even invoking random diversity, there are some diseases which morphologically over-lap others (see p. 96). The histological features mimic those of another. The resemblance is often only superficial, but sufficient to cause confusion if the appearances are interpreted solely by pattern recognition. Critical assessment, of the heuristic type, is usually much more reliable in these instances.

In practice, many histological diagnoses are made by a combination of heuristic analysis and pattern recognition. Commonly, pattern recognition tells one that a lesion is a malignant neoplasm, but then one must go on to ask consciously, "Is this carcinoma, sarcoma, lymphoma, etc.?" "Is this primary or metastatic?" Experienced histopathologists use pattern recognition more frequently than do trainees, but even so it is prudent to test the validity of the diagnosis by systematic analysis and then recheck the diagnosis after the report has been written. For example, an experienced histopathologist may judge a lymph node biopsy to show reactive follicular hyper-plasia after only 15 seconds' examination by recognising the histological pattern; but then the histopathologist should check heuristically the validity of that interpreta-tion by asking specific questions ("Do the follicles possess a mantle zone?" "Do the follicles show normal polarity?" etc.) so that the possibility of follicular lymphoma is actively excluded.

The analogy with art seems to be stronger than with, say, mathematics, though the gradients that connect art and sciences are possibly continuous.

"One gradient, for instance, leads through the so-called exact sciences like chemistry through biochemistry to biology, then through medicine—which is, alas, a much less exact science—to psychology, through anthropology to history, through biography to the biographical novel, and so on into the abyss of pure fiction. As we move along the sloping curve, the dimension of "objective verifiability" is seen to diminish steadily, and the intuitive or aesthetic dimension to increase" (Koestler 1969).

Clinicopathological Integration

A recurrent problem is the extent to which the pathologist should be influenced by clinical information when making a histopathological diagnosis. Sometimes, when no clinical information is provided, this does not arise. But because there is a tendency to read the expected into a histological picture automatically, at the expense of overlooking the unexpected, some pathologists find it helpful to examine the section or gross appearances before seeing the clinical data. This initial unbiased opinion may need revision with subsequent knowledge of the clinical context, but this approach at least has the merit of giving each diagnostic problem a totally fresh appraisal.

Where an unequivocal diagnosis can be made on the morphology alone by virtue of some pathognomonic feature (e.g. acid-fast bacilli in a caseating granuloma denote tuberculosis) then clinical information, though welcome, is superfluous. That a clinician suspected lymphoma rather than tuberculosis does not alter the interpretation of the histological appearances. But often the diagnosis cannot be made on the morphology alone and the appearances must then be interpreted with knowledge of the clinical context. The danger here is that erroneous clinical suspicions about the diagnosis, detailed on the request form, may be assimilated by the histopathologist when interpreting the histology; if the clinician incorrectly assumes that the biopsy report reflects an opinion reached on purely histological grounds, then the erroneous clinical suspicions will be reinforced, i.e. "double-counting" the clinical suspicions (Schwartz et al. 1981). The solution adopted by many histopathologists is to have the clinical information available, but to first look at the histology before reading the request form.

Ultimately, in most cases clinical data is essential for meaningful interpretation of biopsy appearances.

Morphology of Disease Processes

Histological appearances cannot be interpreted unless one is familiar with the cardinal morphological features of disease processes.

Evolution of Lesions

Most textbook descriptions of the histology of diseases tend to concentrate on the fully evolved lesion unmodified by therapy. However, the different patterns produced by the same disease at different stages are well illustrated by the evolution of lesions in skin (Ackerman and Ragaz 1984). Time, in this context, can be regarded as the fourth dimension (Pinkus 1960). In bullous pemphigoid, for example, the typical lesion is a subepidermal bulla. As the bulla heals, the epithelium regenerates over the floor and the bullous cavity now appears to be intraepidermal. Finally, the remains of the bulla are seen as a small subcorneal cleft containing inflammatory cells and debris.

Fixed histological images are the "still life" studies of disease. From a static picture the nature of the disease must be deduced, just like trying to identify the theme of a stage play from one or two still photographs. The changing pattern of a disease with time is analogous to seasonal changes in plants or metamorphosis in insects: an oak tree can be identified from its buds, leaves, flowers or acorns, even from the bark if the "biopsy" is taken from the trunk; the insect can be identified from its egg, larva, pupa or imago. So it should be possible, with experience, to identify diseases according to whether the lesions are early, typical or late. The question rarely arises with tumours because they generally maintain their character throughout their history, but many inflammatory and reactive processes undergo structural changes with time. Similarly, the disease may be in an active or quiescent phase (e.g. ulcerative colitis) with somewhat different histological features in each instance.

Extrapolation from the histological appearances often results in a predictive designation of a lesion as, for example, "premalignant" or "premycotic" or, conversely, "postnecrotic" or "resolving".

If the evolution of a disease is arrested by therapy or interrupted by some natural reaction the result is often an atypical picture ("forme fruste"). Alternatively, a lesion can be biopsied before the pathognomonic features have appeared.

Modification by Treatment

The otherwise typical histology of a lesion may be modified by therapy in such a way that the biopsy appearances can be misinterpreted. Radiotherapy, for example, may induce regression of a tumour but excite bizarre pseudomalignant reactive changes in adjacent connective tissue (Fig. 5.4). Application of podophyllin to warts induces mitotic abnormalities that appear deceptively malignant (Sullivan and King 1947).

It is rare today to see the classical changes of primary thyrotoxicosis in thyroidectomy resections because of the standard practice of giving pre-operative treatment with drugs such as iodine and carbimazole. Similarly, in patients with chronic inflammatory disease who have been treated with steroids the lesions may be atypical and the diagnosis difficult to establish histologically once therapy has been started.

Pruritic skin lesions may be modified simply through the patient scratching them.

Significance of Specific Morphological Features

Rather than attempt to give a comprehensive synopsis of the pathognomonic features of specific diseases, I feel it would be more useful to summarise the significance of

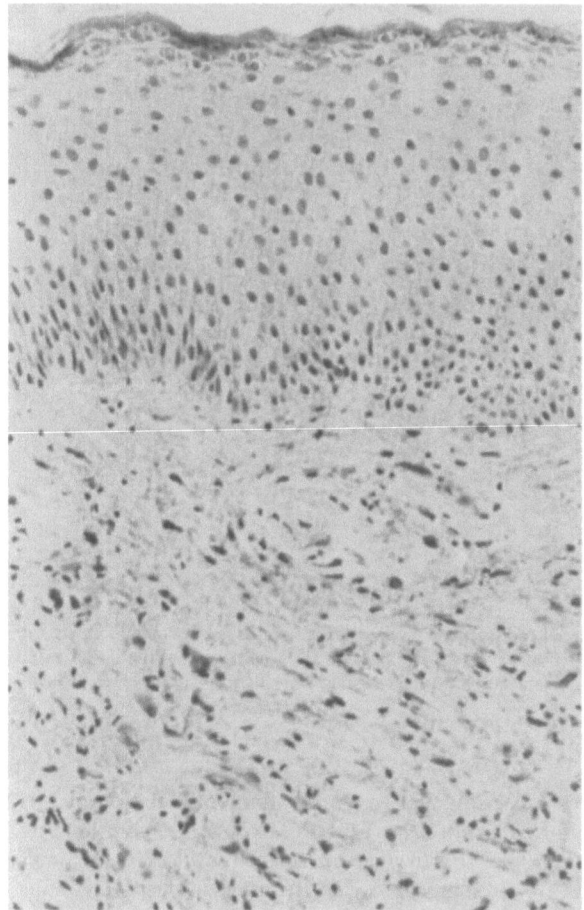

Fig. 5.4. Bizarre dermal fibroblasts, simulating the pleomorphism and hyperchromaticism of neoplasia, at the site of a basal cell carcinoma treated by radiotherapy. Haematoxylin and eosin, ×215

certain histological appearances and comment on their diagnostic usefulness. It is possible to give only the broadest hints as to the conclusions which may be drawn and to point to misleading features which, though histologically prominent, may distract from the real diagnosis. In all cases the diagnosis is made on the entire picture rather than on isolated features. Extensive use of monographs, atlases and textbooks is to be strongly encouraged.

Non-inflammatory Tissue Reactions. Primary changes in a tissue often evoke secondary reactions in an adjacent tissue or cell population. Stromal reactions to tumours, proliferative responses to local damage and metaplastic reactions are commonly seen.

Desmoplastic reactions to tumours are characterised by the formation of a dense collagen-rich fibrous stroma, as exemplified by the scirrhous variant of infiltrating ductal adenocarcinoma of the breast. The infiltrating neoplastic cells of lobular mammary carcinomas may be so sparsely distributed in the desmoplastic stroma that they

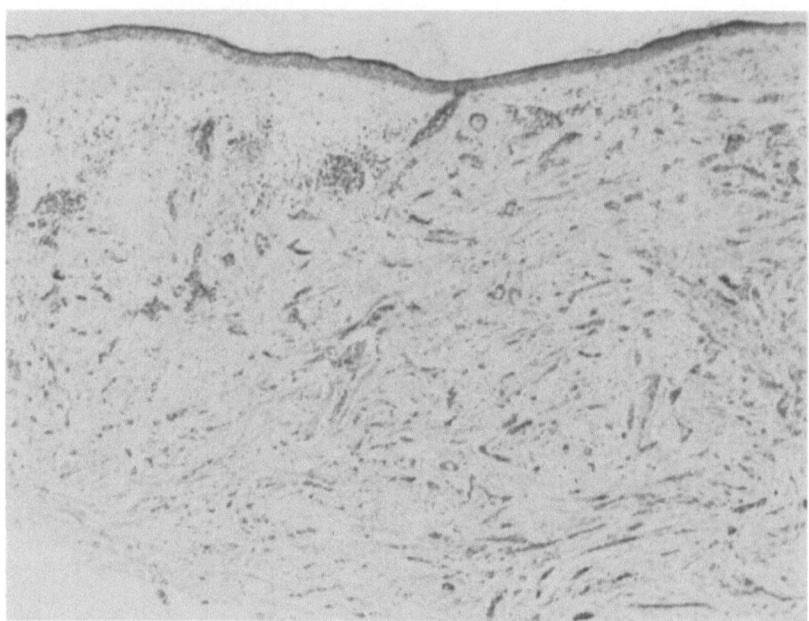

Fig. 5.5. Morphoea-like variant of basal-cell carcinoma. The thin strands of neoplastic cells are associated with a marked dermal desmoplastic reaction. At low magnification the true nature of the lesion could easily be unrecognised because the actual neoplastic component is relatively inconspicuous. Haematoxylin and eosin, ×215

may be overlooked on a cursory inspection. In any unexplained fibrotic lesion it is worth considering the possibility of a desmoplastic reaction to a diffusely infiltrating tumour. Further sections may be required to demonstrate the primary disease process conclusively and stains for mucins are often useful to identify otherwise inconspicuous adenocarcinoma cells. Other desmoplastic tumours include the morphoea-like variant of basal-cell carcinoma of skin (Fig. 5.5) and the desmoplastic type of malignant melanoma. Elastosis is a feature of some breast carcinomas; tumours rich in stromal elastic fibres are more likely to be oestrogen receptor positive.

The ovarian stroma is a specialised endocrine tissue. Luteinisation may be occasionally induced by metastatic carcinomas, notably those from the large bowel.

Metaplastic bone is a rare feature in tumour stroma, but transitional-cell carcinomas of bladder seem to have a particular propensity to induce this reaction.

Dense stromal infiltration by lymphocytes, plasma cells and macrophages is associated with improved prognosis in many tumours (Underwood 1974) and is such a characteristic feature of some (e.g. seminoma, medullary carcinoma of breast) that it may aid diagnosis (Fig. 5.6).

Proliferative reactions in non-neoplastic lesions may be a consequence of local tissue damage. Glomerular injury often results in cellular proliferation and accurate recognition of such changes has considerable importance in the diagnosis and classification of glomerulonephritis. In the liver, marked bile ductular proliferation in primary biliary cirrhosis is probably an abortive attempt to reconstitute the biliary passages following interlobular duct destruction, though this reaction is also seen in several other hepatic disorders.

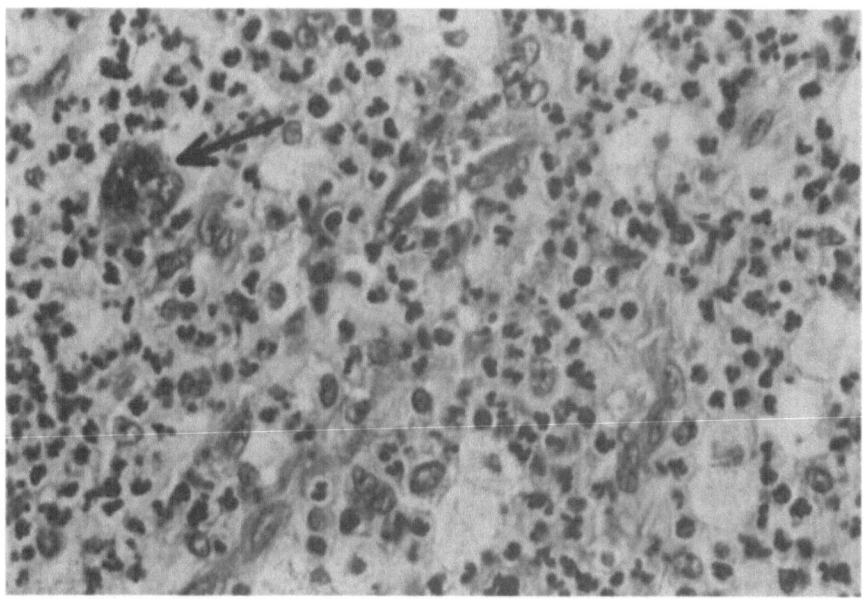

Fig. 5.6. The inflammatory variant of malignant fibrous histiocytoma in which the pleomorphic neoplastic histiocytes (*arrowed*) are admixed with numerous neutrophil polymorphs. This can simulate an inflammatory process and result in failure to recognise the lesion as neoplastic. Haematoxylin and eosin, × 558

Liver-cell proliferation is an inevitable response to hepatocellular necrosis. Regeneration in architecturally intact lobules can result in restoration of the liver to its pristine state, whereas if the lobular architecture has been dissected by bridging or confluent necrosis there is a tendency for asymmetric regeneration and nodule formation leading to cirrhosis.

Crypt hyperplasia is a common sequel of villous damage in, for example, coeliac disease. This distinguishes the villous shortening seen in coeliac disease from the atrophy of villi in malnutrition, etc. which tends to be characterised by crypt hypoplasia.

Mesothelial surfaces of the peritoneum and pleura can exhibit reactions to local damage so florid that they can be mistaken for neoplastic proliferation. Synovium reacts to local injury by undergoing hyperplasia, which eventually gives rise to a villous configuration. Synovial biopsies showing only villous changes with chronic inflammatory infiltration raise the possibility of rheumatoid disease, but those features alone are insufficiently specific to justify a certain diagnosis.

Epidermal reactions are diverse, but their recognition is a useful adjunct to the interpretation of skin biopsies. Actively proliferating epidermis is characterised by acanthosis, increased mitoses, absence of stratum granulosum and abundant parakeratosis. This is designated a "psoriasiform" reaction because of its resemblance to the *disease* of psoriasis. "Lichenoid" reactions, as seen in lichen planus among other conditions, denote diminished epidermal turnover, morphologically characterised by few mitoses and prominence of the stratum granulosum. Pseudocarcinomatous epidermal hyperplasia is seen in association with a variety of disorders, ranging from some fungal infections to granular-cell tumours.

Local damage can induce changes in the differentiated character of epithelial surfaces. Mild degrees of dysplasia are not uncommon within or adjacent to inflamed lesions. The mitotic activity, nuclear hyperchromaticism, and pleomorphism of regenerating epithelium at the edge of ulcers can simulate preneoplastic dysplasia. Metaplastic reactions of diagnostic importance include Hürthle-cell metaplasia in Hashimoto's disease, intestinal metaplasia in association with chronic gastritis and carcinoma, and pseudopyloric metaplasia in the small bowel in Crohn's disease. Squamous metaplasia in the prostate is often seen at the edge of ischaemic lesions.

Inflammatory Reactions. The fluid component of acute inflammatory exudates is difficult to see histologically except when the protein content is sufficiently high to render it eosinophilic. Alternatively, oedema may be deduced from the presence of clear areas of separation between cells or fibres (e.g. spongiosis).

The significance of cellular exudates rests upon identification of the predominant cell type. Neutrophil polymorphs suggest a reaction to either pyogenic organisms or tissue necrosis. Eosinophils suggest allergic reactions, fungal or protozoal infections, Hodgkin's disease and eosinophilic granuloma. The type of inflammatory reaction dictates the special stains that should be applied to elucidate the cause.

Lymphocyte and plasma-cell infiltrates usually indicate chronic inflammation, but very dense pure infiltrates sometimes raise the possibilities of lymphoma or myeloma. Immunohistological methods can be used to investigate this by staining for kappa and lambda light chains; neoplastic infiltrates are predominantly monoclonal. Dense perivascular infiltrates of plasma cells raise the possibility of syphilis, with or without evidence of gummatous necrosis.

Mast cells have attracted relatively little diagnostic significance, probably because they are indistinct in routine haematoxylin and eosin stained sections. Notable exceptions are, of course, urticaria pigmentosa and rare instances of mast-cell neoplasia.

Macrophages are also difficult to see in haematoxylin and eosin stained sections unless they have exercised their phagocytic potential and ingested some sort of particulate material. Macrophages with foamy cytoplasm stuffed with fat are referred to as xanthoma cells. Sometimes such cells are multinucleate and the fat is concentrated in a foamy cytoplasmic rim around the centrally located nuclei; these "Touton" giant cells are often seen in histiocytic tumours (e.g. fibrous histiocytoma). Other macrophage-derived giant cells may be of Langhans' type with peripherally located nuclei, classically but not exclusively seen in tuberculosis, or foreign-body cells with randomly located nuclei in reactions to particulate material (Fig. 5.7). The associated histological features are much more important than the detailed nuclear conformation in making a diagnosis.

Granulomas, defined as aggregates of macrophages or macrophage-derived cells, are seen in a variety of conditions. Giant cells are often but not always present and should not be used as a morphological marker for granulomas when scanning a section at low or medium power. The granulomas in mucosal biopsies of Crohn's disease in particular tend to be small, often devoid of giant cells, and consist of a loosely knit collection of epithelioid cells. Granulomas in mucosal biopsies of the distal gastrointestinal tract can be a reaction to extravasated faecal material or mucin from damaged glands. Acid-fast bacilli, fungi or parasites should always be searched for if these infections could explain the histological or clinical state. Other agents which elicit granulomas include zirconium and beryllium. Sarcoid granulomas occur both in the disease of sarcoidosis and as a localised reaction to lesions such as tumours. Granulomas in the liver close to degenerate interlobular ducts suggest primary biliary

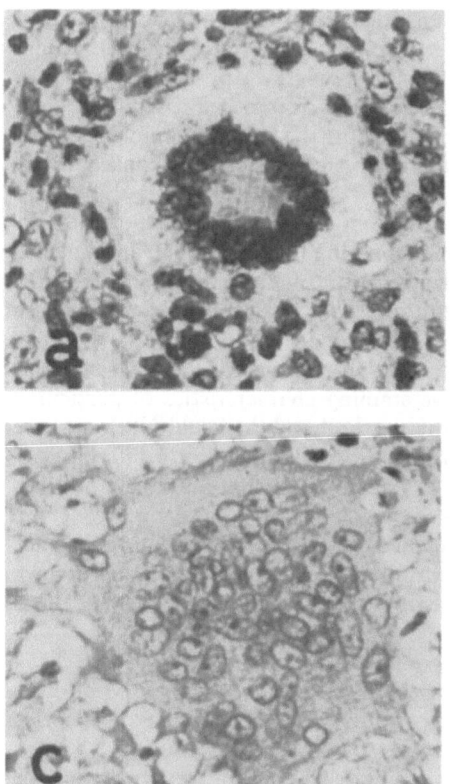
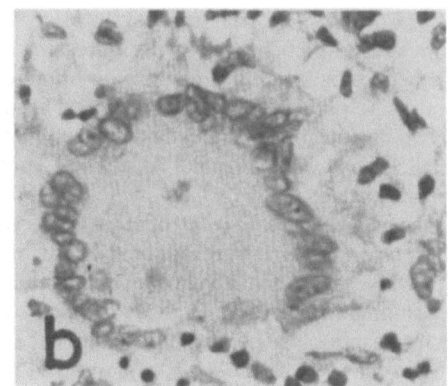

Fig. 5.7a–c. Giant cells derived from macrophages. a Touton cell with central nuclei and peripherally vacuolated cytoplasm in a fibrous histiocytoma. b Langhans' cell with peripheral nuclei in a tuberculous lesion. c Randomly dispersed nuclei in a giant cell in a reaction to silica particles embedded in skin ("pseudotuberculosis silicoticum"). Haematoxylin and eosin, × 558

cirrhosis. Some drugs, notably phenylbutazone, may give granulomatous reactions in the liver. Hepatic granulomas are common in sarcoidosis and are found in some patients with tuberculosis. However, the causes of hepatic granulomas are protean and they occasionally remain unexplained in an individual case.

The main worry with granulomas is the possibility of missing a tuberculous infection. Negative Ziehl-Neelsen staining of tissue sections never absolutely excludes tuberculosis because the organisms are usually very sparse and if the histological context is right the diagnosis must either be accepted and treated or confirmation awaited from the more sensitive microbiological methods.

Blood vessels play an active part in inflammatory reactions and may eventually become damaged or destroyed in the process. Sometimes the inflammatory reaction seems to be located actually in the wall of a blood vessel, i.e. vasculitis. The term "vasculitis" usually needs qualification according to the type of vessel involved (arteritis, capillaritis, venulitis) and the type of tissue reaction (fibrinoid necrosis, leukoclastic, giant cell). Elastic stains may be necessary to identify vessels that are extensively damaged.

Reactive lymph nodes should be specifically assessed for T-zone or B-zone activity. Expansion of the deep cortex with prominent post-capillary venules favours a T-dependent reaction, whereas large active germinal centres and medullary expansion indicate B-dependent activity.

Micro-organisms. Atlases of parasitology are an indispensable aid to the identification of metazoa and protozoa; with increasing international mobility, tropical infestations are becoming much more common outside endemic areas. If the parasite is indistinct in routinely stained haematoxylin and eosin sections then it may be missed. Amoebae are an important example of this trap. Wrongly ascribing the lesions of amoebic colitis to ulcerative colitis results in delayed specific treatment with serious consequences for the patient.

Fungi likewise may be difficult to see in routinely stained sections, but they do stain strongly with periodic acid-Schiff or methenamine (hexamine) silver reagents. For those who seek a succinct aid to the identification of fungi in tissues, I would recommend the practical guide devised by Anthony (1973). Size, shape and type of tissue reaction are helpful pointers to the type of fungus.

Gram's stain is the most widely used general stain for bacteria. It must be done with care to preserve the positive or negative staining characteristics of particular bacteria. However not all bacteria stain by this method and the Ziehl-Neelsen stain is required to demonstrate *Mycobacterium tuberculosis*, Wade-Fite (or Fite-Faraco) method for *Mycobacterium leprae*, and the Levaditi or Warthin-Starry method for spirochaetes (Chap. 3).

Intracellular inclusions may betray the presence of a virus infection and the exact type of inclusion may assist in precise identification (see below). Electron microscopy is necessary to see the virus itself; the ultrastructural features are a further guide to classification.

Intracellular Inclusions. Inclusions visible only by electron microscopy are not dealt with here, nor are ingested organisms or phagocytic debris (e.g. Flemming's corpuscles in germinal centres).

Table 5.2 gives the significance, definite or possible, of a range of the more common and important inclusions (Fig. 5.8).

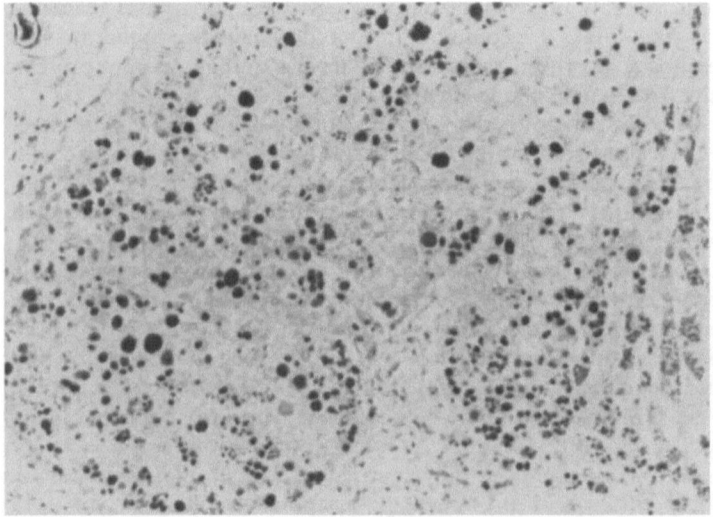

Fig. 5.8. Intracytoplasmic globular hyaline inclusions in hepatocytes in a-1-antitrypsin deficiency. Liver biopsy. Periodic acid-Schiff after diastase treatment of section, × 155

Table 5.2. Intracellular inclusions of diagnostic value

A. Cytoplasmic	Location	Morphology	Significance
i) *Viral*			
Measles	Giant cells	Eosinophilic[a]	Measles
Negri	Neurones	Eosinophilic	Rabies
Guarnieri	Epidermis	Eosinophilic	Smallpox
Molluscum	Epidermis	Eosinophilic	Molluscum contagiosum
Halberstaedter-Prowazek	Conjunctiva	Purple (Giemsa)	Trachoma, etc.
ii) *Non-viral*			
Russell (and other globular hyaline inclusions)	Plasma cells	Eosinophilic, globular	Immunoglobulin retention
	Hepatocytes	Eosinophilic; globular	Probable α-1-antitrypsin deficiency
	Tumour cells	Eosinophilic, globular	Notably yolk-sac tumours and liver-cell carcinomas
Michaelis-Gutmann	Macrophages, urinary tract	Calcified, laminated	Malakoplakia
Spironolactone	Adrenal cortex	Eosinophilic, globular	Spironolactone treatment
Mallory	Hepatocytes	Eosinophilic, reticular	Alcohol, Indian childhood cirrhosis etc.
Hamazaki-Wesenberg	Lymph node, sinus-lining cells	Brown, elongated	Unknown
Schaumann	Giant cells	Basophilic, laminated	Sarcoidosis, sarcoid reactions and non-sarcoid granulomas
Reinke	Interstitial gonadal cells	Eosinophilic, crystalloid	Interstitial-cell tumour
"Private" acini	Tumour cells	AB/PAS "target"	Adenocarcinoma, usually lobular carcinoma of breast

B. Nuclear	Location	Morphology	Significance
i) *Viral*			
Cytomegalovirus	Usually epithelium	Eosinophilic	Cytomegalovirus infection
Measles	Giant cells	Eosinophilic	Measles
Chickenpox	Epidermis	Eosinophilic	Chickenpox
Herpes simplex	Epidermis, etc.	Eosinophilic	Herpes simplex infection
ii) *Non-viral*			
Lead	Liver, renal tubules	Basophilic, globular	Lead poisoning
Glycogen	Hepatocytes	PAS positive, globular	Miscellaneous, including Wilson's disease
Dutcher-Fahey	Plasma cells	Eosinophilic, globular	Macroglobulinaemia

[a] Most viral inclusions are phloxine positive in the phloxine/tartrazine stain.

Patterns of Degeneration and Necrosis. Cellular degenerations like cloudy swelling and fatty change reflect serious metabolic disturbances and yet, taken in isolation, they may have very limited diagnostic specificity. Fat stains on frozen sections are useful in the identification of xanthomatous lesions and lipid-rich tumours (e.g. lipid-rich carcinoma of the breast, liposarcomas, ovarian thecomas), but the mere presence of fat is alone insufficient evidence of adipose differentiation.

Intracellular hyaline inclusions are listed elsewhere in this chapter. Mallory's ("alcoholic") hyalin in hepatocytes is a distinctive type of intracellular degeneration seen in active alcoholic liver disease and a range of other hepatic disorders, including Indian childhood cirrhosis and Wilson's disease. Hepatocytes harbouring hepatitis virus B particles often have "ground glass" cytoplasm; this can be confirmed by the Shikata orcein stain, aldehyde fuchsin, or by an immunohistological method for HB_sAg.

Extracellular degenerations include those that are grouped under the purely descriptive name of "hyaline". Hyaline is, of course, not a single substance but a description of the amorphous appearance shared by amyloid, arteriolar lesions in diabetes mellitus and benign phase hypertension, dense sclerotic collagen, elastosis in skin, "toxic" hyalin around vessels in granuloma faciale, hyaline alveolar membranes in certain pulmonary conditions and thickened basement membranes. Precise characterisation of a hyaline material requires special stains or electron microscopy.

Calcification of soft tissues may be dystrophic, as a result of local tissue damage (e.g. fat necrosis), or "metastatic", as in a patient with hypercalcaemia. Even if some local cause is evident it is worth considering the possibility of hypercalcaemia as a permissive factor. Calcification of elastic fibres is suggestive of pseudoxanthoma elasticum. Calcified microspherules (psammoma bodies) are a feature of the stroma of papillary adenocarcinomas, notably of thyroid and ovary, and are also seen within the cellular whorls of meningiomas. Concretions of iron and calcium salts in fibrous nodules in the spleen, named Gamna-Gandy bodies, are a feature of congestive splenomegaly. Deposits of calcium salts in tissues are easy to recognise because they are strongly haematoxyphilic.

Apart from the structural character of necrotic lesions (coagulative, caseous, liquefactive, etc.), diagnostic importance is attached to their precise distribution in a tissue or organ. The word "apoptosis" has been coined to describe death or drop-out of individual cells, as seen in some relatively slowly growing tumours with a paradoxically high mitotic rate and in involuting organs. Another example is the acidophilic Councilman body of acute hepatitis.

Indeed, in the liver, necrosis patterns have considerable diagnostic and prognostic significance. The term "piecemeal" necrosis denotes degeneration and death of groups of periportal hepatocytes resulting in erosion of the limiting plate. This is a feature of chronic aggressive (active) hepatitis. Bands of necrosis or fibrosis linking portal tracts, or central veins and portal tracts, is known as "bridging", This severe architectural disturbance signifies likely progression to cirrhosis. Confluent necrosis is seen in fulminant acute liver damage.

In zonal necrosis the changes are confined to some anatomically defined region of an organ or organ subunit; this usually has a pathophysiological explanation. Examples include centrilobular hepatic necrosis in severe cardiac failure or in toxic damage such as paracetamol poisoning, and papillary and cortical necrosis in the kidney. Generally speaking, the identification of zonal necrosis in an organ should make one seek toxic or haemodynamic explanations.

"Fibrinoid" necrosis, like hyaline, is a purely descriptive term. Fibrin deposition, collagen degeneration, and immune-complex-mediated damage, alone or in combination, can give the histological picture of fibrinoid necrosis. It can be seen in vasculitis (e.g. Arthus reactions or polyarteritis nodosa), malignant phase hypertension, and necrobiotic nodules (e.g. rheumatoid disease, granuloma annulare, necrobiosis lipoidica).

Some histological features may point to previous local damage when active lesions are absent at the time of biopsy. These include deposits of haemosiderin, either free or as haemosiderin-laden macrophages, and remodelling or distortion of the elastic fibre or reticulin framework of the tissue.

Pigments and Minerals. Exogenous and endogenous pigments can be useful markers of certain disease processes. Usually their innate colour and location is sufficient for identification, but it may be necessary to resort to specific staining or bleaching reactions (Table 5.1).

Lipofuscins are brown pigments seen as perinuclear granules in atrophic tissues ("wear and tear"). They are acid-fast by the long Ziehl-Neelsen method. A lipofuscin pigment is present in macrophages of the colorectal lamina propria in melanosis coli. Ceroid, a related pigment, is seen as diatase-resistant PAS-positive material in Kupffer cells in resolving hepatic damage.

Melanin, confirmable by the Masson-Fontana stain, may be found not only in melanocytes, normal and neoplastic, but in macrophages that have ingested it ("melanophages") and in epidermal basal cells to which it has been donated by melanocytes. The presence of melanin in a cell therefore does not necessarily mean that the cell has synthesised it.

Haemosiderin is haemoglobin derived, has a brilliant golden-yellow granular appearance, and stains with Perls' Prussian blue reaction. The precise location is often helpful in distinguishing haemosiderin deposition due to trauma, haemorrhage, haemolysis, or inflammation from that due to defective iron metabolism (haemochromatosis). Formalin and mercury salts produce artifactual pigments during fixation. The former, like malaria pigment and the pigment associated with *Schistosoma mansoni*, can be removed with alcoholic picric acid. The black pigment of ochronosis is polymerised homogentisic acid.

Exogenous pigments include carbon, commonly seen in the lung and its regional lymph nodes, and metals, such as silver in basement membranes in argyria.

Crystalline Particles. Crystals, like pigments, may be either endogenous or exogenous. Identification is made by size, shape, location, type of tissue reaction and behaviour in polarised light. The identification of crystals in pathological specimens is comprehensively dealt with by Johnson (1972); without resorting to specialised crystallographic techniques, it is possible to identify most of the crystalline material likely to be found in biopsies and resections (Table 5.3).

In addition to identification from the physico-chemical properties of the crystal itself, the type of tissue reaction should give a rough indication of the nature of the material. A giant-cell reaction is typically elicited by the aggregates of needle-shaped urate-crystals (Fig. 5.9). Talc, starch and silica also stimulate a foreign-body reaction. Silica implanted in skin produces a granulomatous reaction that can mimic tuberculosis (so-called pseudotuberculosis silicoticum); the fibrogenic effects of silica in the lung are well known. Asbestos inhalation leads to pulmonary fibrosis and also has oncogenic effects, notably producing carcinoma of the lung and pleural and peritoneal mesothelioma. Thorium dioxide (thorotrast) is now used rarely, though

Table 5.3. Identification and diagnostic significance of crystals

Substance	Crystal morphology	Polarised light	Staining reactions	Significance
Apatite	White, irregular	Isotropic	Haematoxyphilic, alizarin red S positive, von Kossa positive	Mineral of bone; "psammoma" bodies in meningiomas, papillary carcinomas
Asbestos	Slender fibres	Variable	Variable Perls' positive coat (i.e. ferruginous bodies)	Asbestos exposure; asbestosis?, carcinoma of lung?, mesothelioma?
Barium sulphate	Pale yellow, granular	Birefringent		Radiological investigations
Calcium oxalate	White, irregular	Birefringent	Alizarin red S negative	Possible oxalosis; normal in thyroid follicles
Calcium pyrophosphate	Rectangular	Birefringent	Alizarin red S negative, von Kossa positive	Pseudo-gout
Cystine	Yellow hexagons	Birefringent	Ferric ferricyanide positive	Possible cystinosis
Protoporphyrin	Acicular	Birefringent	Fluorescent	Porphyria
Silica	Variable	Birefringent		Silica dust exposure; silicosis?
Starch	Spherical	"Maltese cross"	PAS positive	Glove powder
Talc	White, irregular	Birefringent		Extraneous; talc granuloma?
Thorium dioxide	Pale brown granules	Isotropic	α-tracks on autoradiography	Thorotrast exposure; hepatic angiosarcoma?, etc.
Tyrosine	Radial groups		Millon positive	Possible tyrosinosis; sometimes in stroma of pleomorphic salivary adenomas
Urates	Yellow needles	Birefringent		Gout

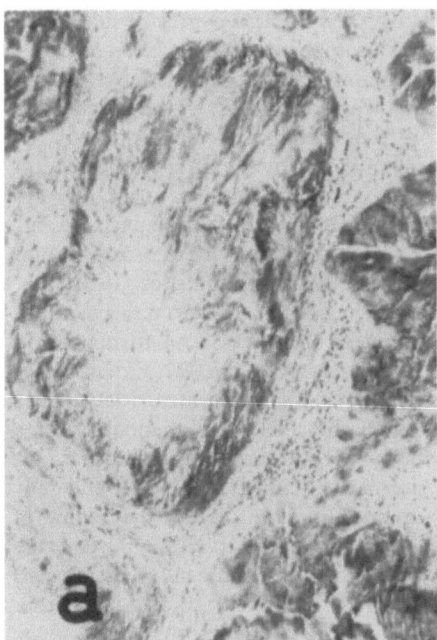

Fig. 5.9a,b. Subcutaneous gouty tophus. **a** Conventional transmitted light microscopy showing large crystalline deposits invested by connective tissue bearing a light inflammatory infiltrate with occasional giant cells. **b** Same field by polarised light; the needle-shaped urate crystals are strongly birefringent. Haematoxylin and eosin, ×65

previous exposure to it should be considered in cases of hepatic angiosarcoma, associated with aggregates of fine and almost colourless granules.

Mitotic Activity. Abnormally frequent mitotic figures may be seen in a tissue for three reasons. There may be either a true increase in the rate of cell division, an increase in the duration of the mitotic phase of the cell cycle, or interruption of the mitotic process (e.g. metaphase arrest due to defective spindle formation). It should not be concluded, therefore, that increased numbers of mitotic figures in fixed histological sections always reflect increased mitotic activity. Genuinely increased mitotic rates imply cellular proliferation, but not necessarily neoplasia.

Mitotic frequency is a much more reliable criterion of malignancy in connective tissue tumours than it is in epithelial tumours.

Abnormal mitotic configurations (tripolar, tetrapolar, etc.) occur commonly in malignant neoplasms, but they are also very occasionally seen in benign lesions with no other features to suggest malignancy.

Differentiation in Tumours. The histogenesis of a tumour is deduced from the presence of distinctive growth patterns (e.g. glands in an adenocarcinoma, vascular channels in an angiosarcoma), some structurally evident cell product (e.g. osteoid in an osteosarcoma, keratin in a squamous-cell carcinoma), or the detection of a cell product by tinctorial, histochemical or immunological methods (e.g. mucin in an adenocarcinoma, immunoglobulin in a plasmacytoma). The degree of differentiation determines the ease with which these features can be perceived or detected.

Among pathologists there is a spread of individual thresholds that define the limits of well- and poorly-differentiated tumours. The practice of most pathologists is to reserve the designation "well-differentiated" or "moderately well-differentiated" for those tumours that are readily recognisable from the growth pattern and general appearance. "Poorly differentiated" or "rather poorly differentiated" tumours are those in which the histogenesis is not immediately recognisable and special stains may be necessary to establish the exact nature of the tumour. "Anaplastic" should be reserved for tumours that show no recognisable differentiated features.

Tumour cells often interact with each other and the cells of the stroma to produce diagnostically useful growth patterns or organoid structures. Gland formation in adenocarcinoma has been alluded to already. Further examples are Call-Exner bodies in granulosa cell tumours, Schiller-Duval bodies in yolk-sac tumours, and rosettes in neuroblastomas and retinoblastomas. Sometimes the cells of an adenocarcinoma are singly dispersed and mucin is retained in the cytoplasm rather than secreted into a glandular space (i.e. "signet ring" cells). Alternatively, each individual cell may be endowed with a "private" acinus lined by microvilli, as in lobular carcinoma of the breast. Organoid changes may be seen in some areas of less differentiated neoplasms. Foci of sebaceous, follicular or sweat-gland differentiation may be seen in some otherwise typical basal cell carcinomas; it may be difficult to decide whether the degree of organoid change justifies designation of the whole lesion as an adnexal tumour.

Another aspect of the question of tumour differentiation is the morphological purity of the lesion. Tumours which show morphological uniformity may be described as "monomorphic". Tumours of variable appearance can be termed "pleomorphic". "Dimorphic" describes tumours which appear to contain two discrete cell populations; examples are sweat-gland tumours where the myoepithelial component participates in the neoplasm, Brenner tumours where circumscribed aggregates of epithelial cells are disposed in neoplastic ovarian stroma, osteoclastomas where osteoclast-like giant cells are arranged in a spindle-cell component, and mesothelial tumours (mesotheliomas, synovial sarcomas, etc.) where neoplastic mesothelium lines clefts in a neoplastic mesenchymal stroma. The unexpected co-existence of two neoplastic cell populations is designated "carcinosarcoma". "Polymorphism" is best illustrated by the multiplicity of differentiated cell lines seen in some teratomas.

If a tumour evades histogenetic classification, or perhaps as an adjunct to histogenetic classification, it may be described in simple structural terms. For this purpose terms such as "villous", "papillary", "follicular", "alveolar", "trabecular", "solid" and "cribriform" have become well established in the language of histopathology (e.g. papillary carcinoma, alveolar rhabdomyosarcoma) (Fig. 5.10).

Fig. 5.10a–p. Tumour growth patterns. a solid (e.g. lymphomas, poorly differentiated carcinomas); b cribriform (e.g. adenoid cystic carcinoma, some mammary intraduct carcinomas); c acinar (e.g. many adenocarcinomas); d follicular (e.g. thyroid); e papillary (e.g. thyroid, some ovarian adenocarcinomas); f alveolar (e.g. melanoma, alveolar rhabdomyosarcoma); g "signet ring" (e.g. some adenocarcinomas); h "rosette" (e.g. neuroblastoma, retinoblastoma); j "dartboard" pattern around ducts (e.g. invasive mammary lobular carcinoma); k trabecular (e.g. some carcinoid tumours); l cellular palisading (e.g. basal-cell carcinoma of skin); m nuclear palisading (e.g. neurilemmoma); n "herring-bone" pattern (e.g. fibrosarcoma); o storiform (e.g. fibrous histiocytoma); p biphasic (e.g. sweat-gland tumours, synovial sarcoma, some mesotheliomas)

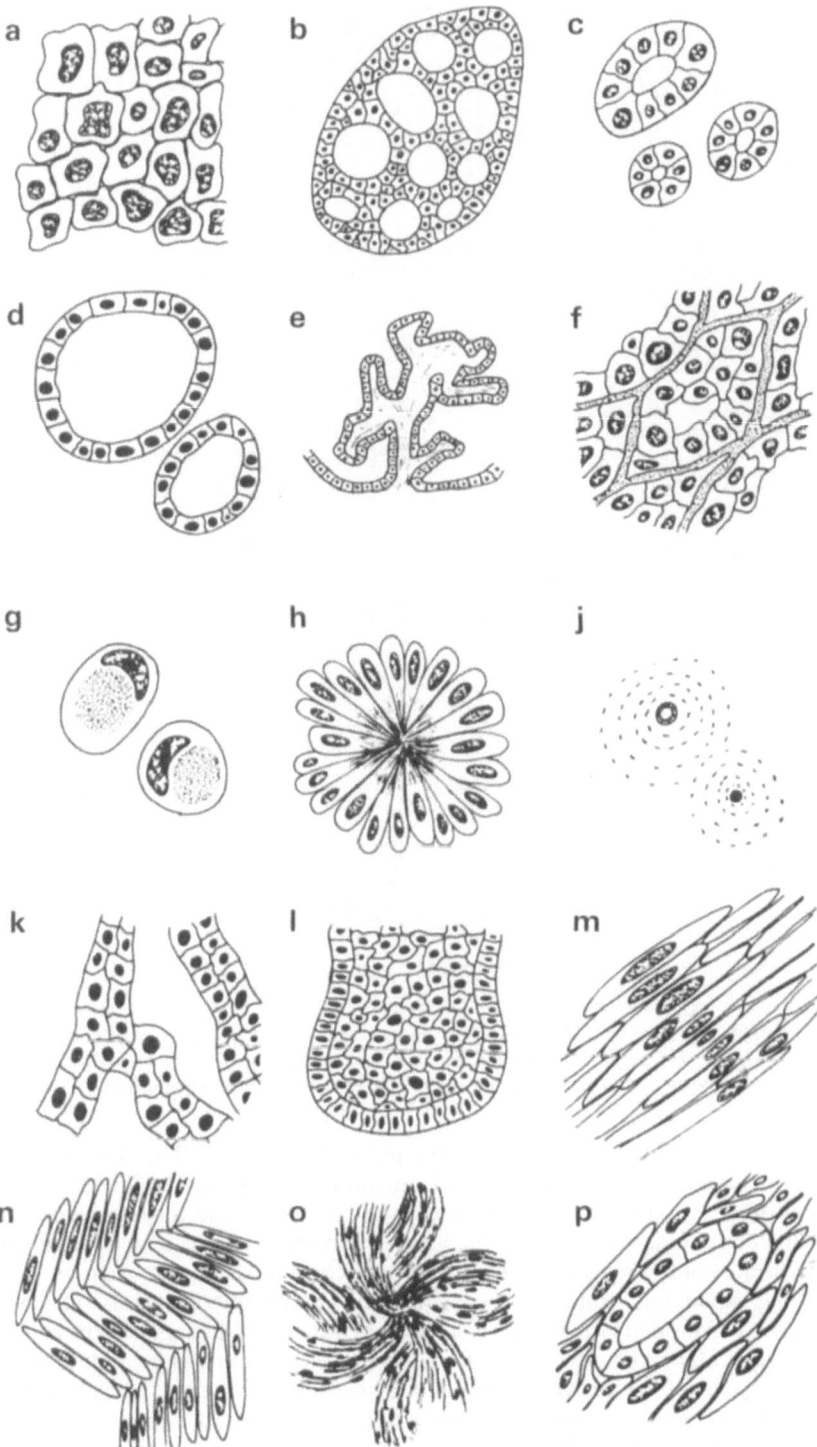

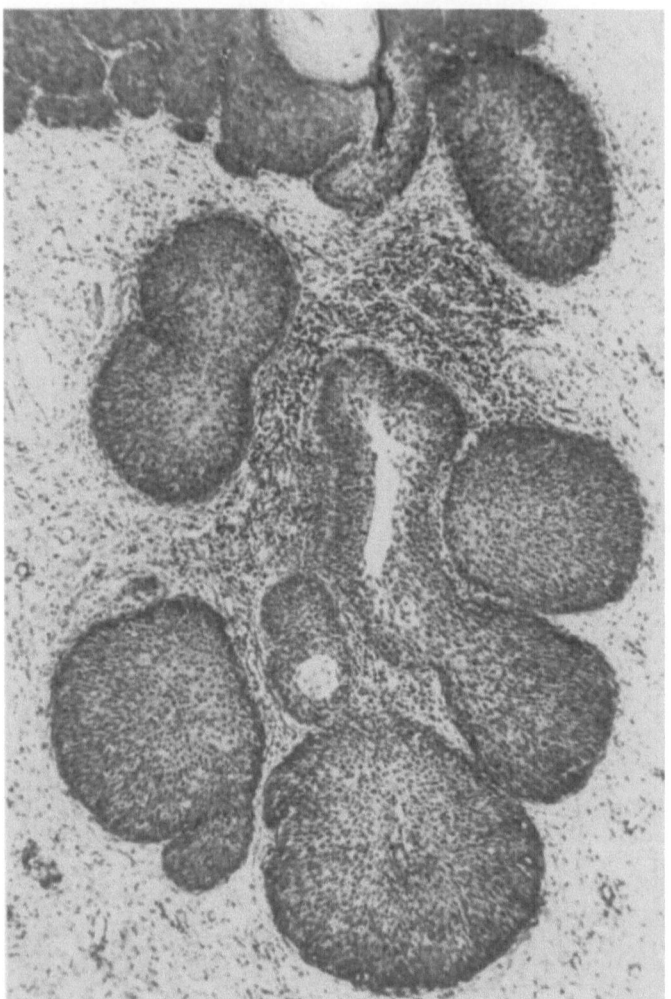

Fig. 5.11. Epithelialisation ("cancerisation") of cervical mucous glands by overlying squamous-cell carcinoma-in-situ. The smoothly delineated islands of neoplastic epithelium contrast with the irregular profiles of cell aggregates seen in true neoplastic invasion. Haematoxylin and eosin, × 65

Identification of a tumour may be hampered by necrosis. The durability of the reticulin framework is remarkable and reticulin staining may well show the basic pattern of the lesion when little is evident on haematoxylin and eosin stained sections.

When doubt exists concerning the local origin of a tumour it may be helpful to demonstrate continuity with some normal structure in the tissue, such as with the overlying squamous epithelium or squamous carcinoma-in-situ in an invasive squamous carcinoma. However, it is sometimes difficult to distinguish true continuity from simple abutment of the two components; deeper sections or additional tissue blocks may, as always, provide more conclusive evidence. The presence of carcinoma-in-situ in an organ containing invasive carcinoma is indeed suggestive, but is not proof of a local origin for the latter.

Disturbances of Tissue Architecture. Disturbances of tissue architecture, even at a cellular level, often result in distinct abnormal morphological patterns. Dysplasia, for example, may be recognised by loss of nuclear polarity within the cell, disturbed cellular orientation, or abnormal cellular maturation.

Minor changes in architecture are often easier to see in sections stained for reticulin. A nodular pattern in hepatic lobules may herald the development of cirrhosis. Effacement of lymph node architecture by lymphomas is often easier to see from the reticulin pattern.

The transgression of normal tissue boundaries by cells is not always due to the active process of neoplastic invasion. Entrapment of epithelium or mesothelium in scar tissue can simulate neoplastic invasion and lead to a mistaken diagnosis of malignancy (p. 133). Evidence of true invasion of vessels and tumour capsule is a prerequisite for the diagnosis of malignancy in certain tumours, notably follicular carcinoma of the thyroid. True invasion must also be distinguished from the less sinister colonisation of glands by non-invasive epithelial tumours growing down ducts; the smooth rounded profile of glands filled by carcinoma contrasts with the irregular angular outline of tongues of genuine invasion into connective tissue (Fig. 5.11). Invasion is never actually seen in sections, it is inferred by the experienced observer from the fixed histological image.

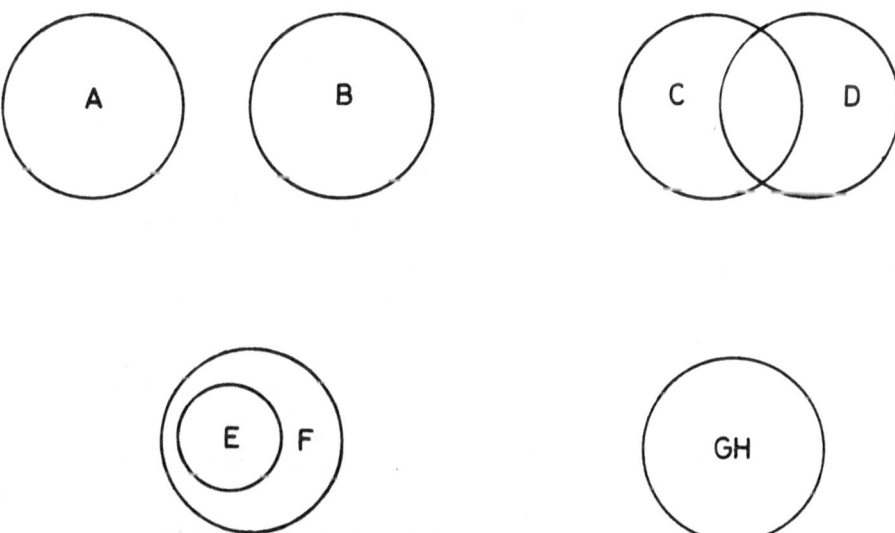

Fig. 5.12. Venn diagrams of possible relationships between the morphological features of different disease entities. Diseases A and B (e.g. ulcerative colitis and colonic diverticular disease) are characterised by distinct sets of morphological features. Diseases C and D (e.g. ulcerative colitis and Crohn's disease of the colon) show a significant degree of morphological overlap; some cases will fall into the indeterminate category. Disease F (e.g. tuberculous lymphadenitis) includes all the morphological features of disease E (e.g. sarcoidosis); additional features characterise disease F (e.g. the presence of acid-fast bacteria). Diseases G and H (e.g. solitary adenomatous polyp of the large bowel and the autosomal dominant condition of polyposis coli) have identical morphological manifestations (a solitary adenomatous polyp resembles a single polyp from a patient with polyposes coli); the disease can be distinguished only by enlarging the scope of inquiry (seeking other lesions, family history, etc.).

Morphological Overlap Between Diseases

If cells could see, communicating with each other in health and disease by visual signals, our morphological approach to the understanding and recognition of disease processes would probably be much easier. As it is, the histopathologist views cells and tissues in a highly artificial way, using stains and other artifacts to amplify and translate the often ambiguous structural manifestations of disease. It is not surprising, therefore, that the morphological message tends to be imprecise, the cardinal features of one disease sometimes resembling or overlapping those of another (Fig. 5.12). At one extreme, we have diseases that display specific pathognomonic features totally exclusive to those particular diseases, and at the other extreme, diseases which morphologically coincide and are therefore indistinguishable by histology alone.

Diseases which cause diagnostic confusion on routine light microscopy alone may be resolved by enlarging the scope of inquiry. Biochemstry, immunology, histochemistry and electron microscopy often give additional information and narrow down the choice of possible interpretations. The probability that the final interpretation will be correct is directly proportional to the amount of information sought.

Finally, we may be faced with the disconcerting possibility that a single disease process will not explain all features of a biopsy. Although in the first instance a serious effort must be made to reconcile apparently disparate histological features by finding a single entity that would explain the whole picture, one may finally be. forced to overcome the natural reluctance to entertain two diagnoses in the same patient (Kopell 1969).

References

Ackerman AB (1978) Histologic diagnosis of inflammatory skin diseases. Lea & Febiger, Philadelphia, pp 157–167

Ackerman AB, Ragaz A (1984) The Lives of Lesions: chronology in dermatopathology. Masson, New York

Anthony PP (1973) A guide to the histological identification of fungi in tissues. J Clin Pathol 26: 828–831

Bartels PH, Weid GL (1981) Automated image analysis in clinical pathology. Am J Clin Pathol 75 [Suppl]: 489–493

Dudley, HAF (1968) Pay off, heuristics, and pattern recognition in the diagnostic process. Lancet II: 723–726

Gombrich EH (1972) Art and illusion, 4th edn. Phaidon, London, p 154

James J (1976) Light microscopic techniques in biology and medicine. Martinus Nijhoff, The Hague

Johnson FB (1972) Crystals in pathologic specimens. Pathol Annu 7: 321–344

Julesz B (1975) Experiments in the visual perception of texture. Sci Am 232: 34–43

Julesz B, Gilbert EN, Shepp CA, Frisch HL (1973) Inability of the human eye to discriminate between textures that agree in second order statistics- revisited. Perception 2: 391–405

Koestler A (1969) The act of creation. Pan, London, p 28

Kopell HP (1969) When one diagnosis won't do then how about two? NY State J Med 69: 2441–2444

Pinkus H (1960) Four-dimensional histopathology. Arch Dermatol 82: 681–698

Schwartz WB, Wolfe HJ, Pauker SG (1981) Pathology and probabilities: a new approach to interpreting and reporting biopsies. N Engl J Med 305: 917–923

Sullivan M, King LS (1947) Effects of resin of podophyllum on normal skin, condylomata acuminata and verrucae vulgares. Arch Dermatol Syphilol 56: 30–45

Thomson SW, Luna LG (1978) An atlas of artifacts encountered in the preparation of microscopic tissue sections. Thomas, Springfield Ill.

Underwood JCE (1974) Lymphoreticular infiltration in human tumours: Prognostic and biological implications. Br J Cancer 30: 538–548

Wallington EA (1979) Artifacts in tissue sections. Med Lab Sci 36: 3–61

Wolman M (1970) On the use of polarized light in pathology. Pathol Annu 5: 381–416

6 Cytology

The history of diagnostic cytology has been reviewed by Bamforth and Osborn (1958) and Spriggs (1977) and is summarised in Chap. 1. Because cytology made fewer demands on technology, it was established earlier than solid tissue biopsies as a diagnostic method. Pioneered by Bennett and his contemporaries in the mid-19th century, cytology then faded into the background in the closing decades of the century as solid tissue biopsies and tissue processing techniques were developed; interest in cytology abated in the face of the apparently greater diagnostic yield and precision of the biopsy. The relative neglect of cytology was reversed dramatically by Papanicolaou's seminal observations in the 1920s on cells exfoliated from carcinoma of the cervix; this marked the renaissance of diagnostic cytology which continues to the present day.

Cytology now has an assured place in clinical practice and is attracting increasing attention because: diagnostic criteria have become more reliably defined enabling more therapeutic decisions to be based on cytology; the technique is now used in conjunction with other investigative methods such as ultrasound and computerized axial tomography enabling previously inaccessible lesions to be sampled; and more precise information can be derived from the very small samples by quantitative morphological methods.

Relative Merits of Cytology and Histology

The relative merits of cytology and histology in their commonest application—the diagnosis of cancer—are summarised in Table 6.1, but it may be misleading to examine the issue in this way because each criterion appears to assume equal weight; this is not so in practice. The cost differences are either immaterial or irrelevant; immaterial because neither technique is a particularly costly aspect of patient care bearing in mind the therapeutic benefit to be derived from a diagnosis, often of early and therefore eminently treatable malignancy, that cannot be established in any other way; irrelevant because the greatest volume of cytology comes from screening programmes of an often asymptomatic population for which biopsies would be inappropriate. The difficulty of performing further investigations such as immunohistochemistry and electron microscopy on cytological samples may be outweighed by the relative ease with which further samples can be obtained for these purposes with relatively little trauma or cost. Cytology is relatively atraumatic, safe, and more acceptable to patients.

Table 6.1. Relative merits of cytology and histology

	Cytology	Histology
Speed	Rapid	Slow
Cost	Low	Relatively high
Accuracy[a]	Fair to good	Good
Precision[b]	Fair to good	Good
Further investigations	Difficult	Easy
Patient acceptability	Good	Often poor
Anaesthetic requirement	None	Usually
Safety	Safe	Variable risks

[a] Lack of false positives and false negatives.
[b] Extent to which a specific or unequivocal diagnosis can be rendered.

Cytology is, of course, an ideal method for population screening, most commonly for cervical cancer and precancer or, for example, urinary cytology for individuals working where there is an increased occupational risk of bladder cancer. Cytology in these situations can generate an increased number of histological preparations because abnormalities discovered in smears may need to be further investigated by solid tissue biopsies before treatment can be justified. So the techniques of cytology and histology are not necessarily mutually exclusive or investigational alternatives.

Cytological Preparations

Cells can be obtained for microscopy by a range of techniques.

Exfoliation

Cells can be scraped from epithelial surfaces by simple abrasion with a spatula (e.g. cervix), brushing (e.g. bronchus), or lavage (e.g. stomach), the choice of technique and instrument being tailored, as in the given examples, to the lesion and organ under investigation. The relatively poor cohesiveness of neoplastic epithelial cells favours their yielding to these methods of sampling. The cells are smeared directly onto a glass slide or brought into suspension, if a brush has been used, by swilling in saline.

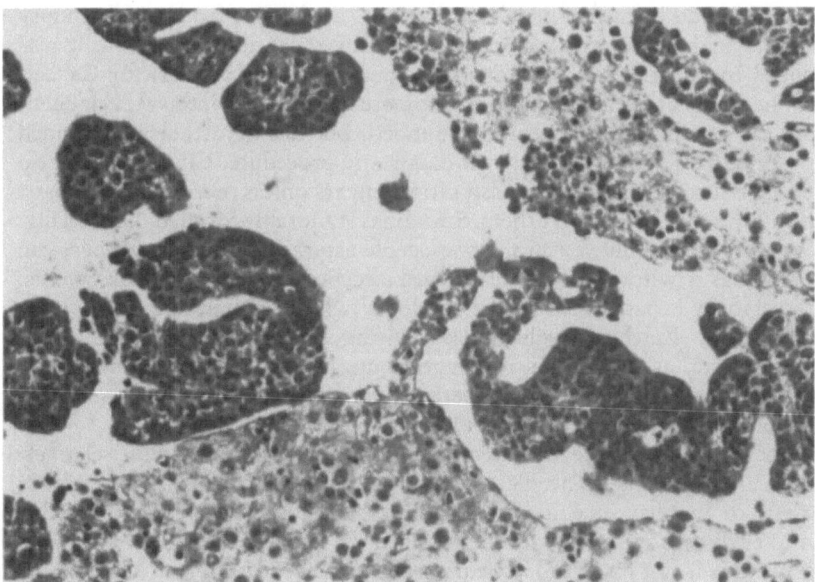

Fig. 6.1. Fibrin clot of a pleural aspirate from a woman with an effusion due to metastatic breast carcinoma. Small solid tumour cell aggregates are present. Haematoxylin and eosin, × 149

Cells from lavage fluids or the washings of brushes can be smeared onto a slide, if necessary after concentration into a pellet by centrifugation, or they can be sedimented directly into a slide using a cytocentrifuge. Brushing or scraping epithelial and mucosal surfaces covers a larger tissue area than biopsy and in many situations is therefore less prone to sampling error than are solid tissue biopsies of selected sites.

Aspiration

Withdrawal of fluid in which cells are naturally suspended is a relatively simple matter and the obvious method of choice for serous effusions in the pleural and peritoneal cavities. It is wise to aspirate the fluid into a bottle or tube containing a small volume of anticoagulant such as heparin; serous effusions rich in fibrinogen (i.e. exudates) often coagulate spontaneously after aspiration and diagnostic cells may become sequestered in the fibrin clot. If this happens inadvertantly, the clot can be processed into a wax block and sectioned for histology; indeed, this has been advocated by some cytologists as an ideal method for examining cells in serous effusions because organoid structures, like glands and papillae, may be seen (Figure 6.1) (De Girolami, 1977a and b; Dekker and Bupp 1978).

Fine-Needle Aspiration Cytology

This is an increasingly popular method for the diagnosis of solid lesions; its history and current status have been reviewed by Frable (1983a,b) and Lever et al. (1985).

Tissue smears made from fine-needle aspirations, as distinct from needle biopsy techniques that yield a solid tissue core, had initially a cautious reception. It was first introduced by Martin and Ellis (1930) at the Memorial Hospital for Cancer, New York. The fear that needle biopsies disseminate tumour cells into vascular channels or along the needle track appears ill founded, but has nevertheless dissuaded many from using this relatively atraumatic diagnostic procedure. Clinicians are also understandably reluctant to undertake aspiration smears unless there is a pathologist available who can reliably interpret them. Scandinavia, notably Sweden, has recently witnessed a resurgence of interest in the fine-needle aspiration biopsy and there can be little doubt that it will eventually find wider acceptance (Fox 1979). In Sweden the aspiration biopsy is performed by the cytologist who interprets the smears; this is commendable but elsewhere might invoke resistance from clinicians who regard direct patient contact as their exclusive right. Some North American studies have shown, however, that the Scandinavian experience can be accomplished in other countries with just as much success (Frable 1976).

A narrow-gauge needle of the sort used for venepuncture if the lesion is relatively superficial, or a spinal-type of needle for more deep-seated lesions, is attached to an empty 10 ml or 20 ml syringe and inserted into the lesion. The plunger is forcibly withdrawn to suck cells out of the lesion and into the needle (a few cells may enter the syringe but this is not essential). The use of a syringe holder (e.g. "Aspir-Gun", Everest Co. Inc., Linden, New Jersey, U.S.A. or "Cameco", Cameco AB, Sweden) enables the operator to exercise more control over aspiration and frees one hand for immobilizing a superficial lesion. To increase the cellular yield and the chances of getting a diagnostic sample the needle is moved in and out of the lesion while suction is maintained. The needle is withdrawn and the contents expressed gently onto glass slides and then smeared and stained. No anaesthetic—local or general—is required, though sedation may be helpful in an anxious patient with a deep-seated lesion. The needle can be targetted to deep-seated lesions by conventional radiology, ultrasound, or computerized axial tomography (Red 1981). Long thin needles can be passed through abdominal viscera to reach retroperitoneal lesions with relative impunity.

Miscellaneous Methods

The scraping technique devised by Dudgeon and Patrick (1927) for the diagnosis of fresh tumour tissue had been revived recently and modified as the "scrimp" (scrape/ imprint) technique (Abrahams 1978). This is claimed to be highly reliable; no false-positives and only one false-negative were encountered in a series of 154 breast lesions. An important advantage is that this sort of method is much less prone to sampling error than frozen sections because numerous levels and facets of the biopsy can be readily assessed.

Stains for Cytology

As with histology the choice of routine stains is largely a matter for institutional, departmental, or individual preferences. However, a summary of the most commonly used stains is here given.

Haematoxylin and Eosin

An obvious choice of stain perhaps, particularly for those who also do much routine histology because the tinctorial characteristics of cells in smears and sections are sufficiently similar (Gubin 1985). But for most cytological work the much-used haematoxylin and eosin stain cannot match the as yet peerless Papanicolaou stain for clear rendition of nuclear detail which is so important for diagnostic interpretation.

Romanovsky Stains

At first sight the May-Grunwald-Giemsa (MGG) stained smears have a seductive beauty. However, the density of the nuclear staining often obscures much of the detail of the chromatin and nucleolar pattern that is so important in the diagnosis of malignancy. The MGG stain is certainly valuable in the diagnosis of haematological problems and lymphoproliferative lesions and it does impart some colour to the acellular background of the smear, such as the characteristic "tiger" background seen in aspirates of seminomas (Lever et al. 1985). It seems to be particularly popular in Europe (Gubin 1985).

Papanicolaou

This has become the favoured stain, particularly for epithelial smears, having first been devised for cervical cytology. Few stains give comparable clarity of nuclear detail. Smears to be stained by this method must have been fixed immediately in alcohol, before the smear dries, for optimum results.

Special Investigations

Immunohistochemistry

Immunohistochemistry can be performed on cytological preparations but the need must be anticipated so that extra smears or imprints can be made while the sample is still fresh. This contrasts with the ease with which solid tissues in wax blocks can be examined retrospectively. Immunohistochemistry is particularly useful for the identification of carcinoma cells in serous effusions, a situation in which reactive but innocent mesothelial cells can commonly masquerade as neoplastic cells; for example, epithelial membrane antigen (EMA) is expressed on the surface of epithelial rather than mesothelial cells, though weak staining of the latter may be seen, and carcinoembryonic antigen expression favours carcinoma rather than neoplastic or reactive mesothelium (Ghosh et al. 1983).

Electron Microscopy

Transmission and scanning electron microscopy are currently under evaluation as diagnostic adjuncts to routine light microscopy in cytology and there is little doubt that these techniques, though time-consuming, lead to more precise and specific diagnoses (Akhtar et al. 1981). For transmission electron microscopy the cells need to be pelleted by centrifugation prior to embedding and sectioning. For scanning electron microscopy the cells must be carefully dried (e.g. by the critical point method) to avoid surface artifacts and then coated with a thin conducting layer, of gold and palladium for example. Because only a relatively small area or number of cells can be examined, both electron microscopic methods are prone to problems of sampling error.

Flow Cytometry

Flow cytometry is an ideal rapid and objective method for the examination of cells in suspension. This method of cellular analysis is described in detail in Chap. 10. pp. 167–170.

Interpretation

Although cytology has many diagnostic applications, its most common use is in the diagnosis of cancer and precancerous lesions. For discussion of the problems of interpretation, we shall concentrate on this specific application since it is here that the greatest difficulties lie.

The features of a cytological preparation that have the greatest diagnostic value in the context of neoplasia and preneoplasia are:

— The cellular yield
— The cohesiveness of the cells
— The nuclear: cytoplasmic ratio
— The nuclear morphology

Generalisations are virtually impossible and can be dangerously misleading, but with this caution the principal differences between benign and malignant cells in cytological preparations are listed in Table 6.2. Some of these criteria are more useful in certain situations than in others. For example, cell yield and cohesiveness are more useful criteria for the assessment of fine-needle aspirates of solid lesions than aspirated cysts or serous effusions. Critical assessment of nuclear morphology is very important, particularly so in serous effusions because it is the only reliable feature which enables reactive and often bizarre mesothelial cells to be distinguished from malignant cells (Fig. 6.2).

The gross pathology is as important as the microscopic appearances and any needle technique gives the person doing the biopsy the opportunity to assess its texture.

Table 6.2. Principal differences between benign and malignant cells in cytological preparations

Feature	Benign	Malignant
Yield	Relatively low	Usually higher
Cohesion	High	Low: disruption of cell clumps occurs at cell/cell interface
Naked nuclei[a]	Often present in smears	Relatively sparse
Nuclear: cytoplasmic ratio	Low	High
Chromatin	Finely dispersed	Relatively coarse and clumped
Nucleoli	Few and small	Often multiple, large and irregular
Background	"Clean"	Often debris and inflammatory cells

[a] Particularly in fine-needle aspirates of breast lesions

Breast carcinomas, for example, have a characteristic gritty texture contrasting with the relative lack of resistance encountered when the needle penetrates benign lesions.

Common interpretative pitfalls include the mimicry of malignant cells by reactive mesothelium, particularly in cirrhosis; reactive epithelial "dysplasias" often associ-

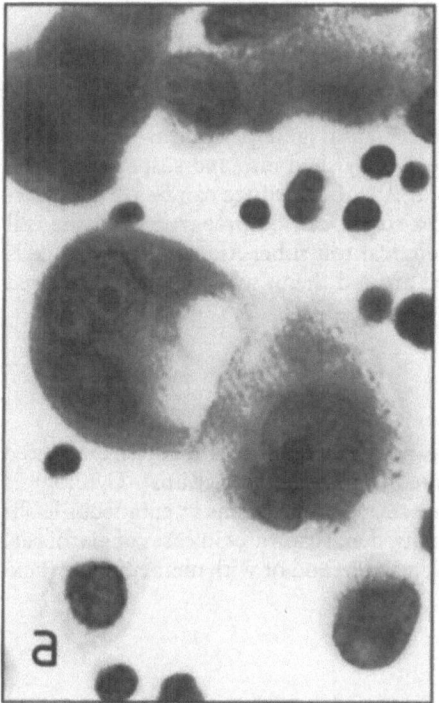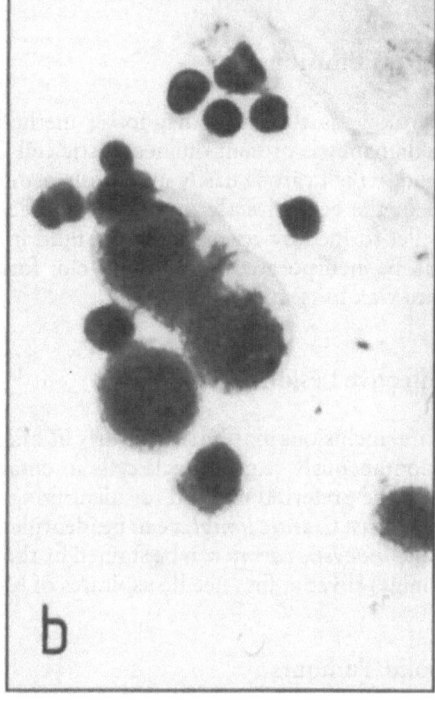

Fig. 6.2a,b. Cytology of benign and malignant serous effusions. **a** Adenocarcinoma cells characterised by large irregular nuclei, fairly coarse chromatin, and multiple nucleoli. **b** Benign effusion comprising lymphocytes and mesothelial cells; the latter have smaller and more regular nucleoli than the adenocarcinoma cells and the peripheral cytoplasm appears vacuolated. Papanicolaou, × 994

ated with inflammation and infections; and atypical cells in the urine of patients on cytotoxic drugs.

Applications

Epithelial "Dysplasias" and Malignancy

This is the most frequent application of cytology and is widely used as a screening procedure, most commonly for cervical neoplasia. The early detection of cervical cancer and precancerous lesions is the subject of major public health screening programmes in many countries. Cytology is the ideal investigative method for these purposes since the affected epithelial surface is accessible and cells can be scraped from the surface with relatively little trauma and certainly no significant morbidity. The large number of cytological smears generated by these screening programmes has activated the development of automatic image-analysis systems to replace the somewhat tedious but unavoidable necessity to examine a relatively large number of normal smears for each abnormal one that is detected. However, although progress is being made there is as yet no automatic system that can entirely replace the expertise of a trained cytologist.

Serous Effusions

Cytology is the ideal diagnostic method for pleural and peritoneal effusions. If malignancy is present the neoplastic cells are usually shed into and suspended in the fluid, which can be easily and safely aspirated. A cytocentrifuge can be used to sediment the cells directly onto a glass slide or a smear can be prepared from the cell pellet formed by centrifuging the fluid in a conical test tube. Alternatively, the cells can be incorporated into a fibrin clot formed around them, which is then processed into wax for sections (Fig. 6.1).

Infective Lesions

Viral inclusions may be seen clearly in smears and the infected cells are often exfoliated spontaneously (e.g. Tzanck cells in cutaneous herpes virus infections). Cytology is also the preferred method for identifying the causative organisms in cutaneous leishmaniasis. *Giardia lamblia* can be identified in duodenal aspirates in cases of giardiasis. *Pneumocystis carinii* can be stained by the Giemsa method or with methenamine (hexamine) silver in fine-needle aspirates of lung.

Solid Tumours

Fine-needle aspiration cytology (FNAC) is now widely used for the diagnosis of superficial and deep suspected tumours (Kline 1981; Frable 1983; Koss et al 1984; Melcher, Linehan and Smith 1984) (Fig. 6.3). It saves the patient anxiety, the surgeon time,

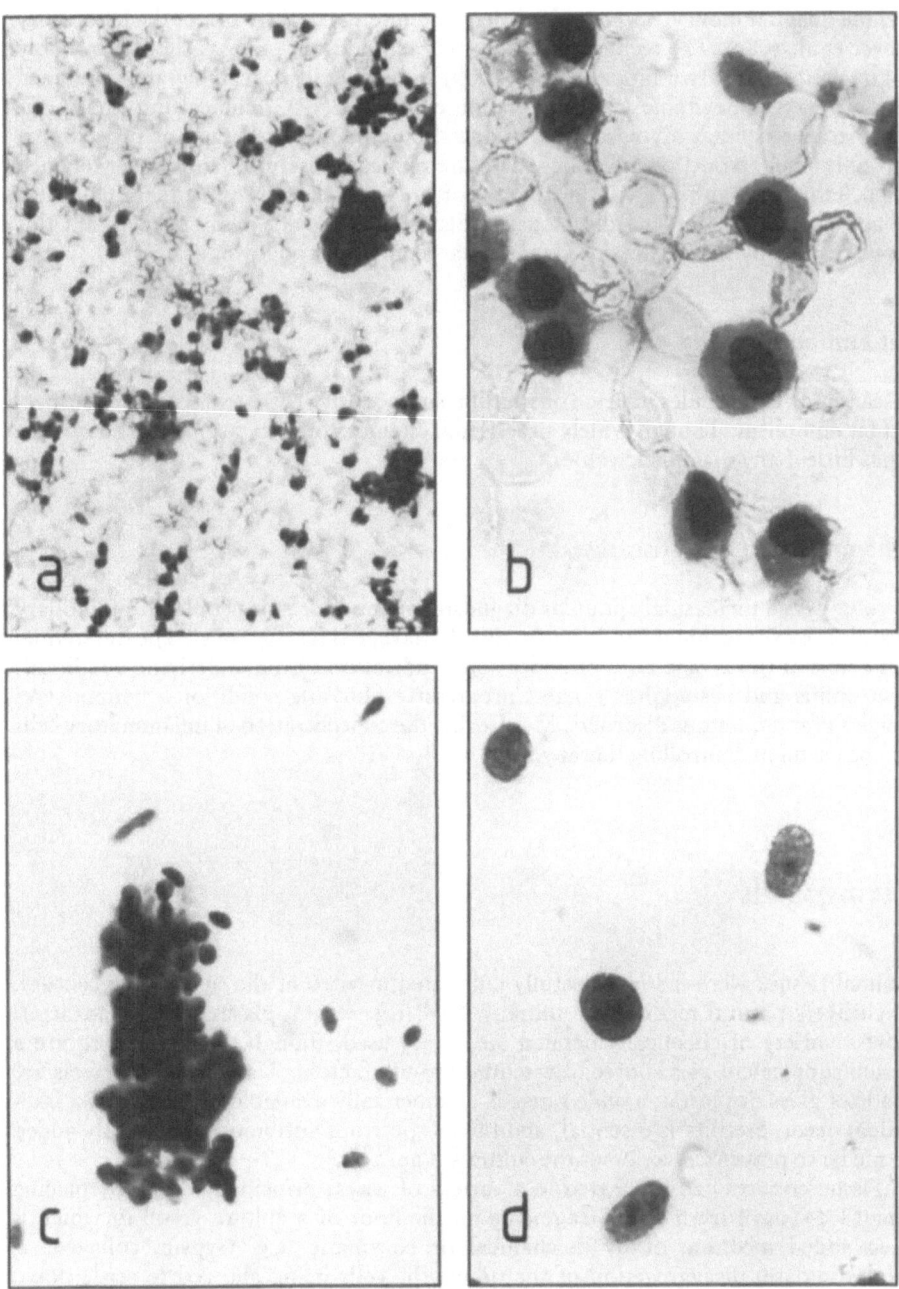

Fig. 6.3a–d. Fine-needle aspiration cytology of a breast carcinoma (**a, b**) and fibroadenoma (**c, d**). Carcinoma: **a** Low-power appearance of typically cellular aspirate from a breast carcinoma in which there are clumps of tumour cells against a background of dispersed intact cells (× 150). **b** Dispersed intact tumour cells have hyperchromatic moulded nuclei (× 994). Fibroadenoma: **c** low-power appearance characterised by clumps of tightly cohesive cells against a background containing "naked" or "stripped" nuclei (× 400); **d** nuclei in background are devoid of cytoplasm and may be of a myoepithelial origin (× 994). (Alcohol-fixed cytocentrifuge preparations stained by Papanicolaou's method.)

and the hospital money, and it adds variety and interest to the work of the laboratory (Lever et al. 1985). The technique does have its limitations: FNAC should probably not be used when larger biopsies, more likely to be diagnostic, can be safely obtained or if surgery is inevitable whatever the outcome of FNAC; clinical information is vital because cellular atypia may be induced by antecedent therapy or procedures; and only unequivocal positive diagnoses are clinically useful (Hajdu and Melamed 1984). However, although in the context of breast cancer FNAC is marginally less reliable than "Tru-cut" needle biopsies (Elston et al. 1978), many believe that this is outweighed by the greater speed, patient acceptability and lower cost.

Fat Embolism

A search for fat globules in urine or sputum is frequently requested in cases of suspected fat embolism. Though widely practised, systematic investigation has shown that it has little if any diagnostic value.

Inflammatory Lung Diseases

Cytology is an increasingly popular diagnostic method for non-neoplastic pulmonary disorders by the technique of bronchoalveolar lavage (Haslam 1984). The differential cell count of the lavage fluid provides some indication of the underlying condition: neutrophils and eosinophils suggest progressive fibrosing conditions; lymphocytes suggest granulomatous disorders. Monitoring the concentration of inflammatory cells can be useful in controlling therapy.

Living Cells

Animal tissues were first successfully cultivated in vitro at the turn of the century. Exclusively natural media were initially used (e.g. serum, plasma, tissue extracts); now a variety of chemically defined media are used, though these often require a serum supplement as a source of essential growth factors. Tissue culture vessels are made of glass or plastic; a wide range is commercially available to satisfy most individual needs. Sterility is essential, and broad-spectrum antibiotics are usually added to media to prevent infection of the cultures (Paul 1970).

Tissue cultures can be started in a variety of ways; principally either by placing small (< 1 mm^3) fresh tissue fragments on the floor of a culture vessel to which is then added medium, or by mechanical or enzymatic (e.g. trypsin, collagenase, hyaluronidase) disaggregation of the tissue, the cells being allowed to settle down in medium on the floor of the vessel. Subcultures are done by enzymatic separation of the cells from each other and the culture surface, washing, and dilution in fresh medium. Most cultures are done at 37°C.

Cultured cells lend themselves to an immense variety of observations, with the advantages offered by tightly controlled conditions, though a major disadvantage is the highly artificial environment. Metabolic requirements can be investigated, cell products assayed, and morphology examined by light microscopy and by transmission and scanning electron microscopy.

Organ cultures are used when it is necessary to preserve tissue architecture or when it would simply be superflous to grow the cells on an artificial surface. Small fragments or thin slices are held on a grid at the interface between the medium and gas phase. This technique has been used in attempts to predict the sensitivity of individual tumours to cytotoxic drugs and to investigate their possible hormone dependence.

Reporting Cytology

A general account of reporting diagnoses is given in Chap. 11. Here we shall concentrate on the problems that are peculiar to cytology reporting.

It is absolutely vital that there is close rapport between the pathologist and clinician and that both understand the implications of what is written on the report and the likely consequences if the report is acted upon.

The pathologist should always err on the side of false negatives rather than false positives; in other words, if in any doubt report as negative and/or suggest a repeat. Clinical action in such cases will vary according to the nature of the problem: for cervical cytology a repeat smear may be more justifiable in the first instance than a solid tissue biopsy; for FNAC of breast lumps, excisional biopsy, perhaps for frozen section diagnosis, would be more appropriate. Reports including the word "suspicious" may precipitate unwarranted clinical action and should be avoided unless each party—pathologist and clinician—clearly understand what is meant in that clinical context.

In negative smears it is important to consider and possibly comment on the adequacy of the sample, otherwise the "negative" report may be misleadingly reassuring to both patient and clinician. For example, a sample sent as "sputum" is inadequate if it is heavily diluted with saliva (recognisable by large numbers of oral squames and bacteria) or contains no pulmonary macrophages (often recognisable by intracytoplasmic carbon granules) to indicate that it is genuinely sputum. To report such samples as "negative" would be misleading because a proper sample has not been obtained. Similarly, an acellular FNAC of a solid lesion suggests that the target has been missed and should not be reported as simply "negative".

Positive (i.e. malignant) reports should be checked and corroborated by colleagues, unless very experienced. It is useful to keep a record of such cases so that subsequent biopsies or resections can be checked against the cytological findings.

Several considerations apply to reporting cervical cytology and various groups, such as the British Society for Clinical Cytology (BSCC) Working Party on Terminology in Gynaecological Cytopathology, have done much to encourage the use of uniform nomenclature and terminology. This is particularly important with cervical cytology which is practised on such a large scale that, if there were significant variations in nomenclature and terminology, it would be impossible to derive any worthwhile national statistical or epidemiological data from the results obtained in different laboratories. The preferred terminology is "cervical intraepithelial neoplasia"—abbreviated to CIN—to describe the spectrum of epithelial abnormalities from the very mild disturbances of differentiation (CIN1) to those seen in smears from severely dysplastic lesions or carcinomas (CIN3). To avoid possible confusion about the significance of CIN3, which can imply either severe dysplasia, carcinoma in situ, or carcinoma in situ with possible invasion, the National Cytology Form HMR 101/5 (1982) used in U.K. laboratories specifies two alternatives for a CIN3 report; either "severe dysplasia/carcinoma-in-situ" or "carcinoma-in-situ/invasive carcinoma".

Spriggs et al. (1978) recommended that the report on a cervical smear should include separate statements or entries, thus:

1. An objective description of the cells present and the abnormalities perceived within them
2. An interpretation of the appearances
3. A recommendation of a course of action (e.g. repeat, biopsy)

Nuclear abnormalities (e.g. hyperchromasia, irregular profile, condensation of chromatin) are very important and largely determine how the smear is reported. The term "dyskaryosis", meaning literally an abnormal nucleus, should be reserved for those nuclear abnormalities which are in excess of those that could be ascribed to inflammation alone; dyskaryosis is therefore reserved for the range of nuclear abnormalities seen in CIN (Spriggs et al. 1978). Although invasive carcinoma cannot be diagnosed with certainty from a smear, there are a few clues which might raise this as a possibility: cells of irregular shape—"fibre cells" and "tadpole cells"; clumps of dyskaryotic cells; fragments of keratinized cytoplasm. The BSCC Working party discourage the use of the term "malignant cells" in a cervical cytology report unless there is unequivocal evidence of an invasive tumour.

Cytopathic changes induced by human papillomavirus may be accompanied by dyskaryosis; management of the patient should be guided by the degree of dyskaryosis (Kaufman et al. 1983).

References

Abrahams C (1978) The "scrimp" technique: a method for the rapid diagnosis of surgical pathology specimens. Histopathology 2:255–266

Akhtar M, Ali MA, Owen EW (1981) Application of electron microscopy in the interpretation of fine-needle aspiration biopsies. Cancer 48: 2458–2463

Bamforth J, Osborn GR (1958) Diagnosis from cells. J Clin Pathol 11: 473–482

De Girolami E (1977a) Applications of plasma-thrombin cell block in diagnostic cytology. I. Female genital and urinary tracts. Pathol Annu [Part 1]: 251–275

De Girolami E (1977b) Applications of plasma-thrombin cell block in diagnostic cytology. II. Digestive and respiratory tracts, breast and effusions. Pathol Annu [Part 2]: 91–110

Dekker A, Bupp PA (1978) Cytology of serous effusions: an investigation into the usefulness of cell blocks versus smears. Am J Clin Pathol 70: 855–860

Dudgeon LS, Patrick C (1927) A new method for the rapid microscopical diagnosis of tumours with an account of 200 cases so examined. Br J Surg 15: 250–261

Elston CW, Cotton RE, Davies CJ, Blamey RW (1978) A comparison of the use of "Tru-cut" needle and fine needle aspiration cytology in the pre-operative diagnosis of carcinoma of the breast. Histopathology 2:239–254

Fox CH (1979) Innovation in medical diagnosis: the Scandinavian curiosity. Lancet I: 1387–1388

Frable WJ (1976) Thin-needle aspiration biopsy: a personal experience with 469 cases. Am J Clin Pathol 65: 168–182

Frable WJ (1983a) Fine-needle aspiration biopsy: a review. Hum Pathol 14:9–28

Frable WJ (1983b) Thin-needle aspiration biopsy. In: Bennington JL (ed) Major problems in pathology. Saunders, Philadelphia

Ghosh AK, Spriggs AI, Taylor-Papadimitriou J, Mason DY (1983) Immunocytochemical staining of cells in pleural and peritoneal effusions with a panel of monoclonal antibodies. J Clin Pathol 36: 1154–1164

Gubin N (1985) Haematoxylin and eosin staining of fine needle aspirate smears. Acta Cytol 29: 648–649

Hajdu SI, Melamed MR (1984) Limitations of aspiration cytology in the diagnosis of primary neoplasms. Acta Cytol 28: 337–345

Haslam PL (1984) Bronchoalveolar lavage. Sem in Resp Med 6: 55–70

Kaufman R, Koss LG, Kurman et al (1983) Statement of caution in the interpretation of papillomavirus-associated lesions of the epithelium of the uterine cervix. Acta Cytol 27: 107–108

Kline TS (1981) Handbook of fine needle aspiration biopsy cytology. Mosby, St Louis Toronto London

Koss LG, Woyke S, Olszewski W (1984) Aspiration biopsy: cytologic interpretation and histologic bases. Igaku-Shoin, New York

Lever JV, Trott Pa, Webb AJ (1985) Fine needle aspiration cytology. J Clin Pathol 38: 1–11

Martin HE, Ellis EB (1930) Biopsy by needle puncture and aspiration. Ann Surg 92: 169–181

Melcher D, Linehan J, Smith R (1984) Practical aspiration cytology. Churchill Livingstone, Edinburgh

Paul J (1970) Cell and tissue culture, 4th edn. Livingstone, Edinburgh London

Red DE (1981) Imaging modalities for biopsy assistance. In: Kline TS (ed) Handbook of fine needle aspiration biopsy cytology. Mosby, St Louis Toronto London, p 8

Spriggs AI (1977) History of cytodiagnosis. J Clin Pathol 30: 1091–1102

Spriggs AI, Butler EB, Evans DMD, Grubb C, Husain OAN, Wachtel GE (1978) Problems of cell nomenclature in cervical cytology smears. J Clin Pathol 31: 1226–1227

7 Borderline Lesions, Pseudomalignancy and Mimicry

Errors in biopsy diagnosis may arise because the morphological features of one disease either overlap or mimic those of another. The practical importance of this point is emphasised in this chapter. A distinction will first be drawn between the concept of borderline lesions, where two or more named conditions (diagnoses) occur within the same continuous disease category, and genuine mimicry, where two or more entirely separate diseases share morphological similarities.

Borderline Lesions

Discrete Versus Continuous Diagnostic Categories

The morphological features of different diseases affecting the same organ are usually so characteristic that they can be used to distinguish one disease from another. Pulmonary tuberculosis can be distinguished from adenocarcinoma of the lung, pemphigus vulgaris from malignant melanoma, colonic diverticular disease from ulcerative colitis, acute tubular necrosis from chronic pyelonephritis, etc. These discrete diagnoses are rarely confused. On the other hand, there are many examples of separately named conditions co-existing within the same broad diagnostic category; the histological distinction between one named condition and another is blurred and often subject to arbitrary criteria. There is no better example than the continuous spectrum from normality to carcinoma, through varying degrees of dysplasia, seen in biopsies and cytological preparations from the squamous epithelium of the uterine cervix. Other similar circumstances include chondroma and chondrosarcoma, the progression from simple fibrosis to nodular regeneration and fully established cirrhosis in the liver, some benign ovarian cystadenomas through borderline lesions to frank cystadenocarcinomas, and the borderline category intermediate between the tuberculoid and lepromatous reactions in the lesions of leprosy.

Use of Criteria

The dysplasia-carcinoma spectrum in the squamous epithelium of the cervix will serve as a model for further discussion of continuous diagnostic categories, and to illustrate the need for criteria so that individual biopsies can be consistently assigned to a named position within the spectrum. Dysplasia, in the present context, strictly refers to epithelial dysplasia and not as loosely applied to, for example, mammary "dysplasia" or fibrous "dysplasia" of bone.

Normal squamous epithelium has an orderly stratified structure. Mitotic activity is confined to the basal layer and the cells progressively differentiate and mature as they move to the surface. The cells within any one zone (basal, stratum granulosum, etc.) closely resemble each other. Squamous-cell carcinoma-in-situ, at the other extreme, is devoid of stratification; mitotic activity and differentiation occur at all levels so that the epithelium now presents a pleomorphic appearance. Between these two extremes, normal and carcinoma-in-situ, there is a spectrum of named conditions (diagnoses)—mild dysplasia, moderate dysplasia, severe dysplasia, referred to in the cervix as CIN (cervical intra-epithelial neoplasia) 1, 2 and 3—the limits of which are, to a certain extent, arbitrarily defined (Fig. 7.1). For accurate cytological or histological diagnosis of lesions in the dysplasia-carcinoma spectrum it is necessary to assess the appearances against a set of previously agreed criteria. Cytologists and histopathologists will resort to their own individual decision rules when judging the

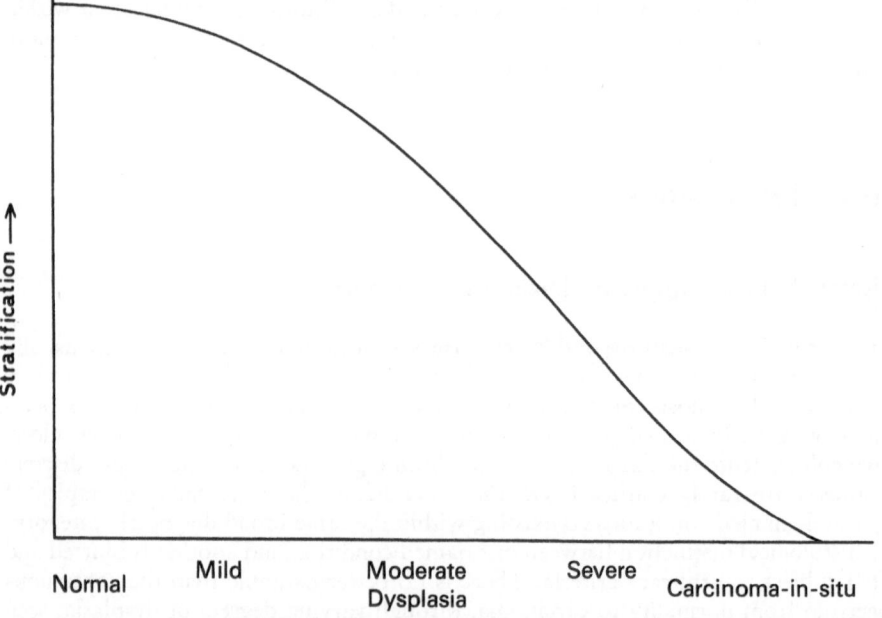

Fig. 7.1. Normality, grades of dysplasia and carcinoma-in-situ gradually merge into each other to form an apparently continuous morphological spectrum. The orderly cellular stratification of normal squamous epithelium progressively declines across the spectrum until it is absent, apart from a trace of surface flattening, in carcinoma-in-situ. The limits of each named category cannot be precisely defined. (Stratification in this context must not be confused with the acquisition of *nuclear* stratification seen in glandular neoplasia)

similarity between the specimen under consideration and the various designated positions in the dysplasia-carcinoma spectrum, the limits of which are set by the criteria (Langley 1978). The reliability of this procedure has been the subject of several studies (Cocker et al. 1968).

Dysplasia in glandular epithelia also merges with carcinoma, for example in the stomach and large bowel. The dysplastic state is evinced by nuclear hyperchromaticism, gradual loss of nuclear polarity (giving the appearance of nuclear "stratification", and the presence of mitotic activity remote from gland crypts (Fig. 7.2). Progression to frank carcinoma, in the absence of visible invasion, is often marked by the formation of "back-to-back" glands, i.e. two or more glandular spaces sharing a party wall without intervening stroma.

A proportion of ovarian tumours, both serous and mucinous (Hart and Norris 1973), can be placed in a borderline category intermediate between the overtly benign and the frankly malignant by assessing features such as stromal invasion and epithelial stratification. Cartilage tumours present a similar problem; a continuous morphological spectrum appears to exist between chondromas and chondrosarcomas. The behaviour of borderline lesions, intermediate between that of unequivocal benign and malignant neoplasms, validates the morphological concept.

Mild dysplasia without premalignant connotations may be seen in inflammatory and regenerative lesions, such as *Trichomonas vaginalis* infection in the cervix and mucosal proliferation at the edge of healing peptic ulcers. Premalignant dysplasia in ulcerative colitis must be distinguished from dysplastic changes seen in areas of active inflammatory disease.

The borderline between what is normal and what indicates disease may also be difficult to define in biopsies. Obvious examples include the skin, where marked differences in the relative proportions of the various epidermal strata can be seen in different sites; a thick stratum corneum would be normal on the sole of the foot but abnormal on the trunk. Mildly abnormal villous patterns in jejunal biopsies from adult Caucasians might be regarded as being within normal limits when seen in biopsies from people of some other ethnic groups or from children (Morson and Dawson 1979).

Pattern recognition is an unreliable method for the assessment of borderline lesions; it is too subjective. These problems should always be analysed in a logical fashion using established criteria (i.e. by heuristic analysis). These criteria are not exclusively histological; they include the site of the lesion, the presence of other disease or lesions, and the sex, race and age of the patient.

Mimicry of Histological Features

Sometimes the histological features of two entirely separate disease entities are so similar that it becomes extremely difficult to distinguish them. One disease mimics another. The degree of morphological overlap may be considerable or slight. Examples include sarcoidosis and tuberculous lymphadenitis, Crohn's colitis and ulcerative colitis, ulcerative colitis and amoebic colitis, mesothelial reactions and neoplastic effusions. If the sets of criteria used to distinguish these conditions are purely morphological, the differential diagnosis often remains difficult to resolve. If

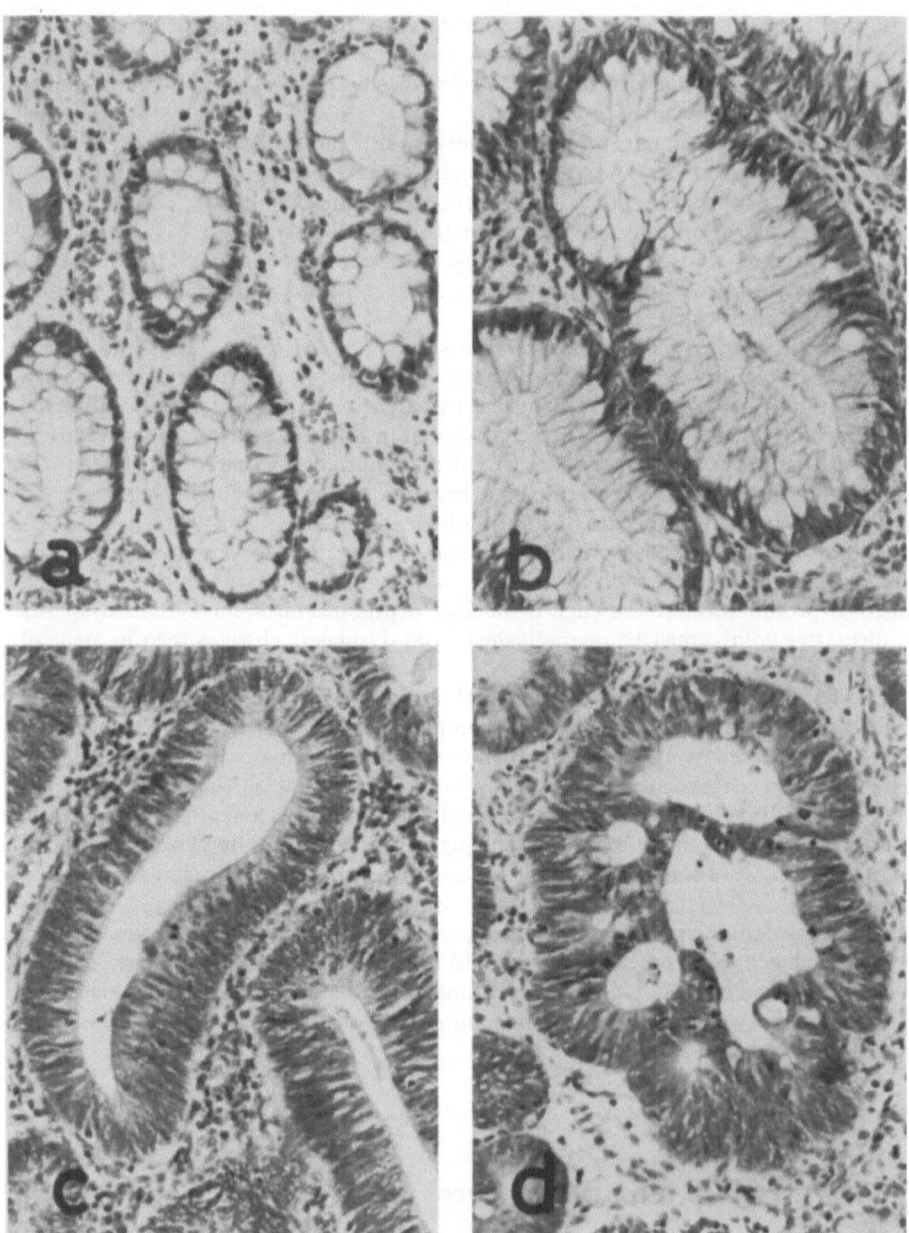

Fig. 7.2a–d. Grades of glandular dysplasia as seen in adenomatous polyps of the large bowel. a Normal glands. Mild b and moderate c grades of dysplasia in which the nuclei are hyperchromatic, crowded, and show progressive loss of polarity. d Severe dysplasia in which adjacent glandular spaces share a common wall of cells ("back-to-back" glands); there are valid reasons for not calling this "carcinoma-in-situ" in the large bowel (see Chap. 11). Haematoxylin and eosin, × 215

the sets are enlarged to encompass clinical, biochemical, microbiological and other criteria, the mimicry may be resolved. But faced with a small specimen and limited information it may be impossible to distinguish two unrelated diseases beyond reasonable doubt. Mimicry is yet another trap for those who rely too heavily on quick diagnosis by pattern recognition. The only way to avoid these potentially serious diagnostic errors is to be aware of the possible dangers and then to analyse the differential diagnostic problem logically.

Pseudomalignancy and Related Diagnostic Problems

Pseudomalignant lesions may seem out of place in an introduction to histopathology, but they are included here to illustrate the concept of mimicry and to show how histological appearances must be carefully analysed if serious diagnostic errors are to be avoided.

Pseudomalignancy is an example of mimicry in which a benign lesion is erroneously interpreted as one that has malignant connotations. Recognition of a lesion as neoplastic, and its further classification as benign or malignant, is based on the use of a set of fundamental pathological criteria. As a general rule, malignant neoplasms are invasive lesions that show excessive mitotic activity, cellular pleomorphism, and nuclear hyperchromaticism, along with features of dedifferentiation. Benign tumours are, in contrast, usually circumscribed lesions and resemble more closely their parent tissue. The identification of benign and malignant tumours is done by recognising and interpreting these gross and histological features. However, the presence of invasion, excessive mitotic activity, pleomorphism and nuclear hyperchromaticism, either alone or in combination, does not always necessarily denote malignancy, and some benign processes produce tissue patterns that will superficially mimic those produced by genuine malignant neoplasms. If the histopathologist is unaware of these conditions, benign tumours and tumour-like lesions can be easily misinterpreted and treated as malignant neoplasms with the risk of serious consequences.

Not only can benign lesions simulate malignant tumours, and vice versa, but also some malignant tumours can simulate those of another type. This is not uncommon with poorly differentiated neoplasms where the lesions are so devoid of distinguishing characteristics that a tumour of one histogenetic type can look very much like that of another (e.g. a poorly differentiated adenocarcinoma lacking glands or mucin, and a poorly differentiated squamous-cell carcinoma lacking keratinisation). Some carcinomas lose their epithelial pattern altogether and acquire a sarcomatoid spindle-cell appearance; squamous-cell carcinomas of the upper respiratory tract seem unusually prone to adopt this sort of disguise (Fig. 7.3; Hyams 1975). Poorly differentiated carcinomas often resemble lymphomas, but may be distinguished by immunohistology or electron microscopy.

Remember also that neoplastic invasion itself is a dynamic process only to be inferred from a static fixed histological image. An aerial photograph of a busy urban scene would not directly indicate whether a person or vehicle was in motion or stationary; the observer infers the presence of movement from the position of the pedestrians or vehicles. A sequence of photographs might show that certain objects had moved to a different position, but the static images do not directly show *how* the movement occurred. So it is that the mere presence of some tissue in an unfamiliar location should not be immediately construed as evidence of neoplastic invasion. The tissue

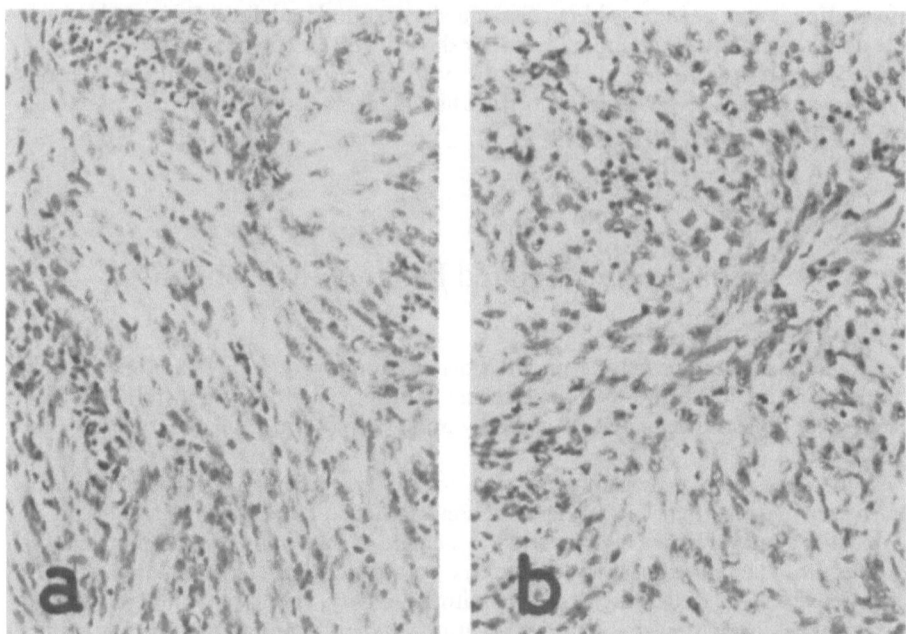

Fig. 7.3a,b. Spindle-cell carcinomas mimicking sarcomas. a Spindle-cell carcinoma of the larynx, a tumour of allegedly squamous histogenesis. b Sarcomatoid renal adenocarcinoma; in other areas the tumour was more typical in its appearance. Haematoxylin and eosin, × 168

may be there as a result of some developmental abnormality (i.e. heterotopia), trauma (i.e. implantation), entrapment (i.e. organisation of inflammatory damage), artifact (i.e. floater or carry-over), as well as, of course, true neoplastic invasion and metastasis.

A common and well-known example will serve as an illustration of pseudomalignancy. Figure 7.4 shows part of a sinister-looking skin lesion from the face in which irregular islands of squamous epithelium invade connective tissue; in other areas there is excessive mitotic activity and pleomorphism. This aggressive initial appearance can give the impression that we are dealing with a well-differentiated invasive squamous-cell carcinoma, an interpretation that is incorrect. These rapidly growing invasive lesions, keratoacanthomas, regress spontaneously. Histologically, a keratoacanthoma can usually be distinguished from a true squamous-cell carcinoma only if, with other subtle features, the crater-like architecture of the former is visualised. A keratoacanthoma is therefore a pseudomalignant lesion.

The converse, though perhaps less common, may also be a problem since malignant neoplasms may superficially masquerade as benign lesions. This can be illustrated by reference to proliferative lesions of mammary duct epithelium. Figure 7.5 shows a mammary duct containing a cribriform arrangement of cells that have a monotonously regular morphology; pleomorphism is not conspicuous. In contrast, Fig. 7.6 shows a similar duct that is filled with a more obviously pleomorphic cell population. Paradoxically, the monotonous appearance of the cells in Fig. 7.5 is more suggestive of malignancy and indeed should be regarded as intraduct carcinoma. In

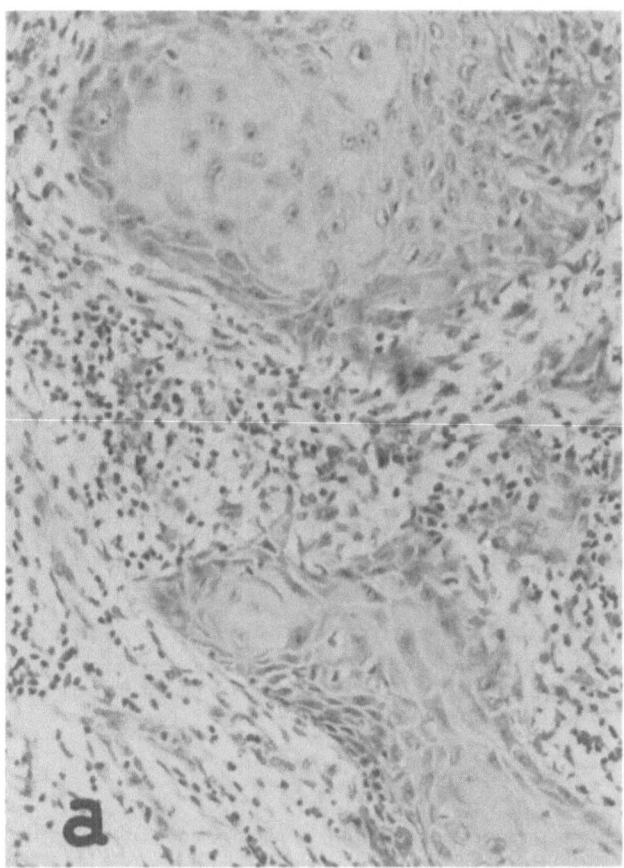

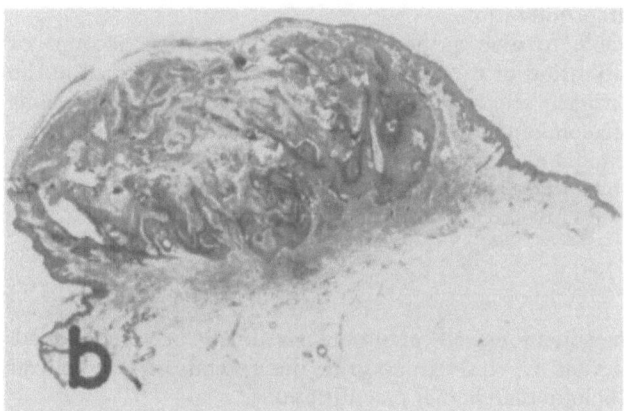

Fig. 7.4a,b. Keratoacanthoma. **a** Irregular islands of squamous cells invading dermal connective tissue and simulating the appearance of squamous-cell carcinoma. Haematoxylin and eosin, × 215. **b** Low magnification reveals the crater-like configuration of the lesion, typical of keratoacanthoma. Haematoxylin and eosin, × 7

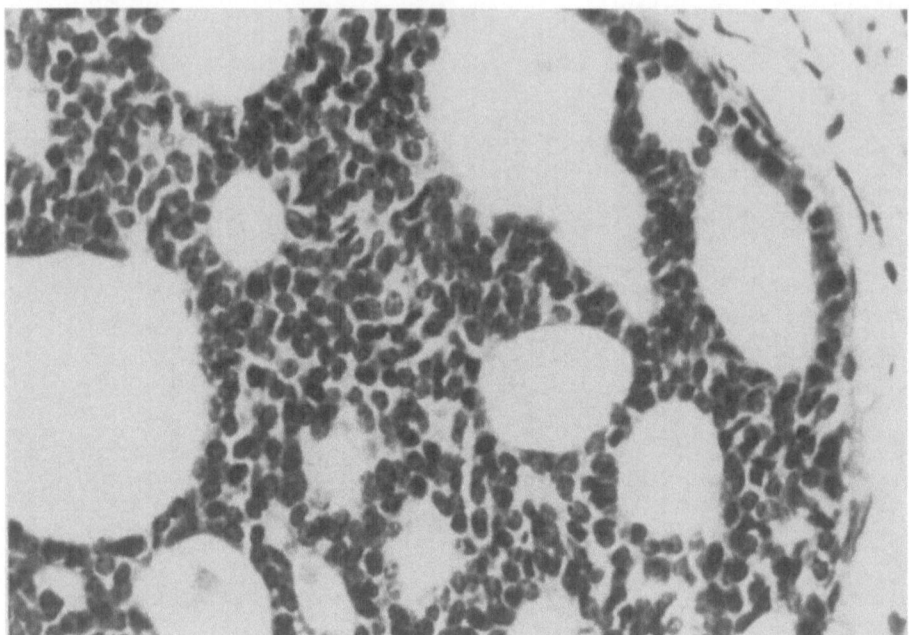

Fig. 7.5. Intraduct carcinoma of the breast. The uniformity of the cells is misleading and might be wrongly interpreted as a benign process. Cellular regularity, nuclear hyperchromaticism and the cribriform pattern favour malignancy. Compare with Fig. 7.6. Haematoxylin and eosin, × 558

the second example, the apparent pleomorphism is due to the presence of a mixture of at least two different cell populations (presumably epithelium and myoepithelium) rather than cellular atypia within a single population of neoplastic cells. Intraduct carcinoma can therefore appear so bland and innocent that it may be dismissed by the inexperienced as a benign proliferation.

Although it is not possible to give a thoroughly comprehensive account of pseudomalignancy, some indication of the extent of the problem is outlined in the following systematically arranged selections. Some additional examples of lesions which create diagnostic confusion, other than pseudomalignancy (e.g. pseudobenign), are also included. For a detailed account of a wide range of problematic lesions I would recommend the text by Park (1980).

Skin

Keratoacanthoma simulating squamous-cell carcinoma has already been mentioned. The architecture of the lesion is vital to accurate interpretation; low-power microscopy is usually more helpful than higher magnifications.

Simple warts are rarely mistaken for carcinoma, but giant condyloma acuminata may be (Davies 1965). Similarly, though basal-cell papilloma (seborrhoeic keratosis) in its pristine state is easily distinguishable from a carcinoma, when irritated or inflamed it can keratinise in its deeper parts and resemble a squamous-cell carcinoma.

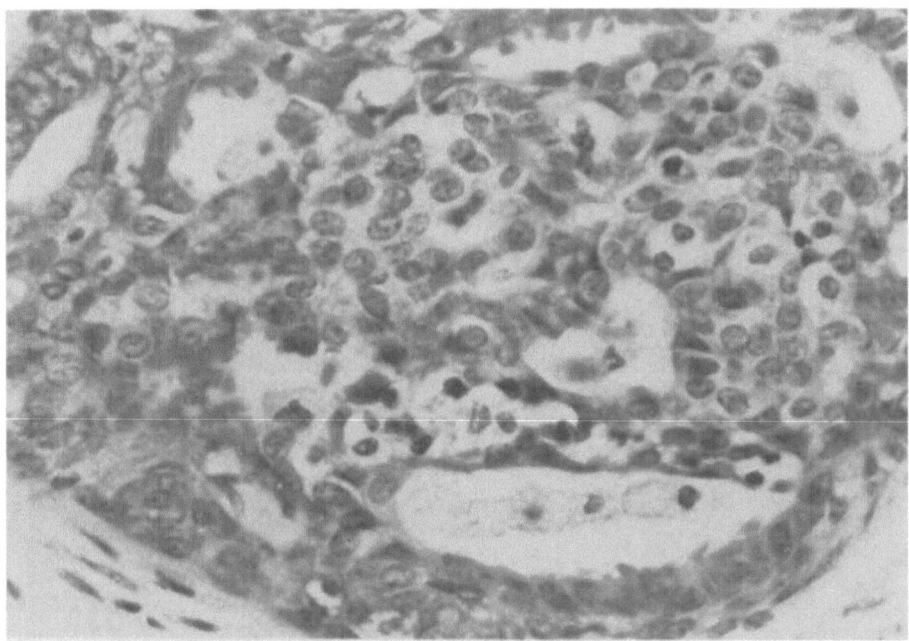

Fig. 7.6. Epitheliosis in a mammary duct. The apparent pleomorphism is due to an admixture of cell types (epithelium, myoepithelium); this pattern may be wrongly interpreted as a malignant process. Lack of cellular uniformity, cellular "streaming", and the irregularity of the glandular spaces favour epitheliosis. Compare with Fig. 7.5. Haematoxylin and eosin, × 558

Among the situations in which pseudocarcinomatous epidermal hyperplasia can occur are at the edge of non-neoplastic ulcers of the skin or, curiously, overlying a granular-cell tumour in the dermis (Fig. 7.7). Absence of pleomorphism is certainly important in establishing the correct diagnosis, but distinction from carcinoma may still be extremely difficult. The context in which the abnormality is seen is just as important as the detailed histology of the epithelial proliferation.

Both carcinoma and cutaneous lymphoma can be simulated by tissue reactions to insect bites (Allen 1948).

Pigment-cell lesions are especially difficult because junctional activity in a benign naevus can both closely mimic malignant melanoma and may also be, in some cases, a borderline condition between a benign lesion and one that is frankly malignant, i.e. "dysplastic" naevus. Knowledge of the patient's age is essential when these lesions are examined by the histopathologist; junctional change is often marked in the growing naevi of prepubertal and pubertal individuals, whereas malignant melanoma is relatively uncommon at this age. Helpful features include maturation of the cells in the deeper parts of the lesions; this is seen in naevi but not, as a rule, in malignant melanomas, though mature cells may be seen in the deeper remnants of a pre-existing naevus that has given rise to a malignant melanoma. However, no single feature can be used to distinguish active naevi from malignant melanomas. Table 7.1 compares some of the features of benign and malignant pigment-cell lesions.

Juvenile melanoma (Spitz naevus) is a benign lesion that can easily be mistaken for malignant melanoma. It is not confined to juveniles. The constituent cells may

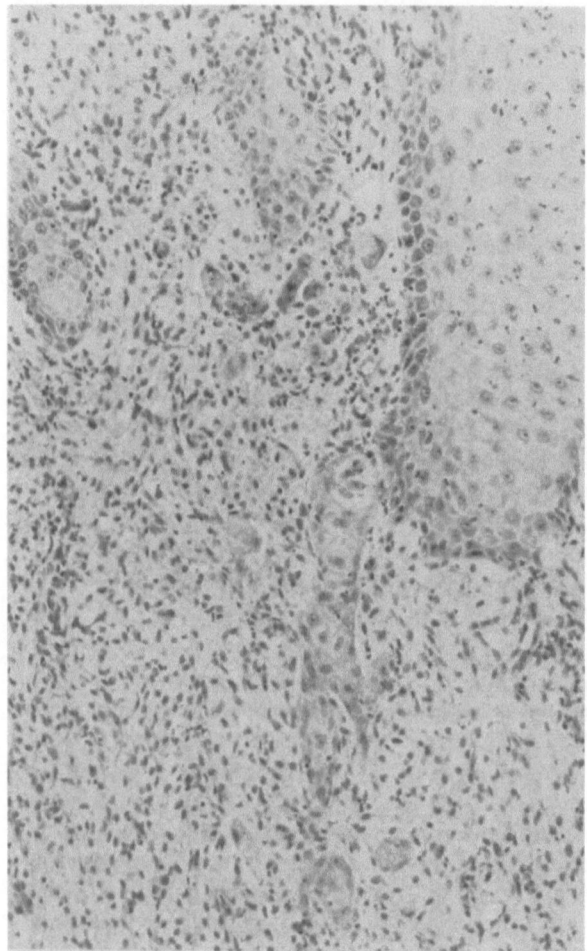

Fig. 7.7. Pseudocarcinomatous hyperplasia of the epidermis over a granular-cell tumour, a not uncommon phenomenon. The irregular squamous downgrowths mimic carcinoma, but can be safely ignored in this context. Haematoxylin and eosin, × 215

Table 7.1. Comparison of benign (naevi) and malignant (melanoma) pigment cell lesions

	Benign	Malignant
Mitoses	Absent or infrequent	Present
Lymphocyte infiltration	Infrequent	Common
Cell maturation in deeper layers	Present	Absent
Cell morphology	Small and regular	Larger and often atypical
Appearance of junctional zone	Uniform size and spacing of junctional aggregates	Irregular junctional aggregates
Cell location	Confined to basal layer of rete ridges	Scattered through epidermis and between rete ridges

be spindle shaped or epithelioid, and some may be multinucleated. The stroma is often oedematous and, superficially, rich in ectatic vessels. Mitoses are common, especially in the superficial component. These are indeed worrying lesions for the histopathologist. Features which suggest malignancy include junctional activity between rete ridges, lack of deep maturation, asymmetry, and no increase in reticulin fibres (McGovern 1976).

Drugs can alter the histological appearances of a benign lesion in such a way that it is misconstrued as malignant. A good example is the induction of atypia and mitotic irregularity in simple warts by the application of podophyllin, which produces a resemblance to squamous-cell carcinoma.

Soft Tissues

Before considering the problem of the histological diagnosis of soft-tissue lesions, it is worth commenting briefly on the macroscopic features of benign and malignant lesions. In most tissues, benign tumours are usually circumscribed growths whereas malignant tumours have poorly defined edges and are infiltrative. The reverse is often the case with soft tissues. Thus, for example, fibrosarcomas tend to be rounded masses and fibromatoses, which are benign proliferative lesions, invariably appear infiltrative.

Under the microscope, to the novice many soft-tissue lesions can be deceptive. Pseudosarcomatous (nodular) fasciitis and proliferative myositis are rapidly growing

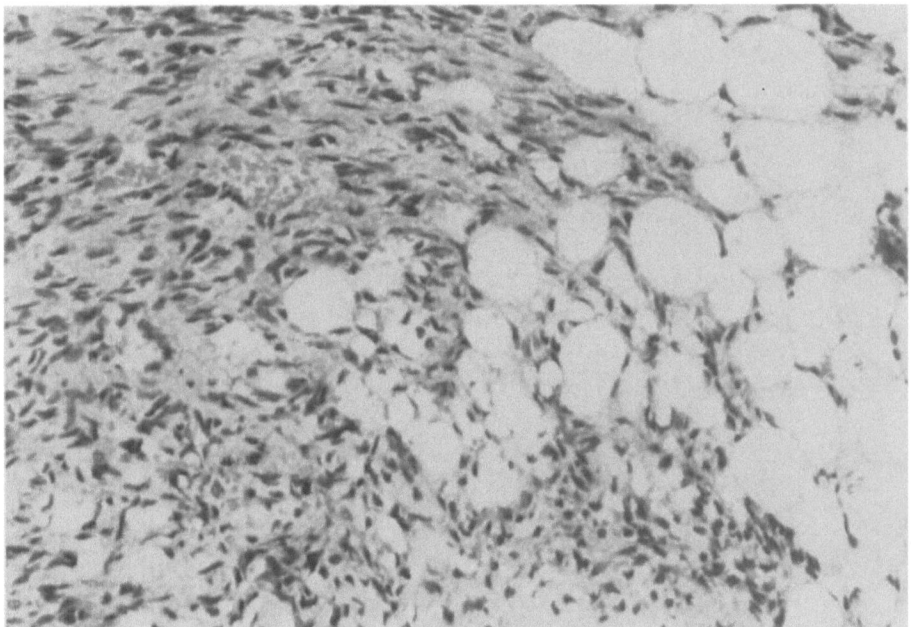

Fig. 7.8. Nodular fasciitis, a pseudosarcomatous benign proliferation of fibroblasts. The lesion is not encapsulated or circumscribed and penetrates the adjacent adipose tissue. Haematoxylin and eosin, × 215

lesions that appear invasive and exhibit mitotic activity. Both can be mistaken for sarcoma (Fig. 7.8).

The greatest number of erroneous differential diagnoses awarded to a single lesion must be claimed by the epithelioid sarcoma (Enzinger 1970). Here the mimicry usually operates in the direction of an undoubted malignant neoplasm that may be wrongly interpreted as being a benign lesion (e.g. nodular tenosynovitis, rheumatoid nodule). Despite the tendency for the lesion to present a somewhat inflammatory histological pattern, epithelioid sarcoma is a malignant neoplasm undoubtedly capable of metastatic behaviour.

Ischaemic and other degenerative changes in benign tumours may cause the neoplastic cells to assume a worrying morphology. The "ancient" schwannoma is a benign neurilemmal tumour in which hypoxia, presumably, leads to cellular enlargement and formation of bizarre nuclei (Fig. 7.9). No mitotic activity is evident in these lesions, a feature that is helpful in distinguishing this degenerative change from the pleomorphism that normally suggests a malignant neoplasm. Similarly, a lipoma can undergo degenerative changes and fibrosis with pleomorphism to give a picture which mimics liposarcoma.

Juvenile rhabdomyosarcoma of the botryoid type presents as a polypoid excrescence on mucosal surfaces, the malignant cells often congregating just beneath the epithelial surface. The bulk of the polyp may contain banal myxoid tissue which misleadingly appears innocent. Hence, the importance of examining the entire lesion with great care cannot be overemphasised.

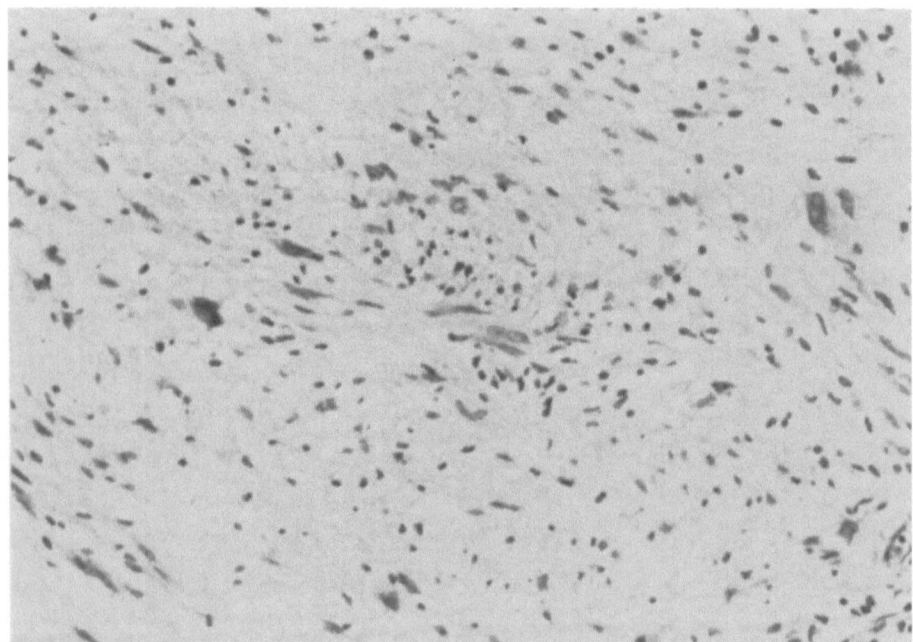

Fig. 7.9. Bizarre atypical nuclei in a degenerate "ancient" schwannoma. These changes are probably a consequence of hypoxia and should not be construed as malignant. Lack of mitotic activity emphasises the benign nature of this process. Haematoxylin and eosin, × 215

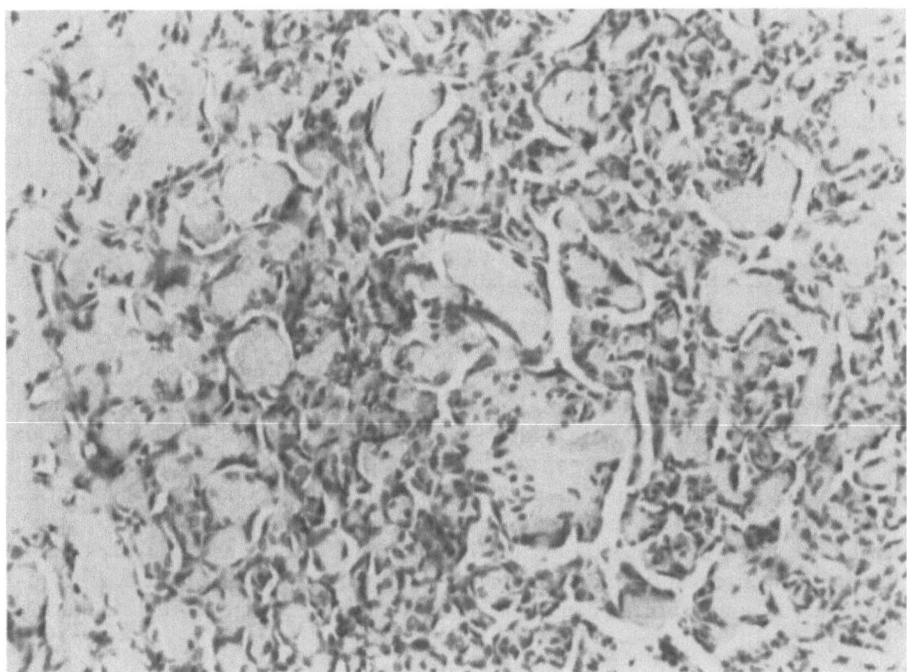

Fig. 7.10. Papillary endothelial hyperplasia ("vegetant intravascular haemangioendothelioma" of Masson). The irregular vascular spaces bordered by plump endothelial cells can lead to this benign lesion being mistaken for angiosarcoma. Haematoxylin and eosin, × 166

Malignant vascular tumours can be mimicked by the exuberant leashes of proliferating small blood vessels within an organising thrombus or haematoma (Fig. 7.10). Also, as pointed out by Stout and Lattes (1967), choriocarcinoma and hypernephroma are tumours that have such a richly vascular stroma that they may sometimes be mistaken for haemangiosarcomas.

Lymphoid Tissues

The histological pattern of nodular follicular centre-cell lymphomas bears such a close similarity to the follicular architecture of normal and reactive lymph nodes that distinction can be extremely difficult; increased awareness of the entity of nodular lymphomas has, in recent years, reduced the frequency of incorrect diagnoses. However, it is worth restating the most useful criteria by which nodular lymphomas and reactive follicular hyperplasia can be distinguished (Table 7.2).

Some specific types of reactive change can simulate malignant lymphoma (Dorfman and Warnke 1974). Saltzstein and Ackerman (1959) first described the occurrence of a lymphadenopathy histologically resembling lymphoma in epileptic patients taking phenytoin. Likewise, infectious mononucleosis and vaccination can both lead to local or even systemic lymph node enlargement, sometimes with architectural effacement, strongly suggesting malignant lymphoma. In one series (Hartsock 1968) 9 out of 20

Table 7.2. Comparison of reactive follicular hyperplasia and nodular (follicular) lymphoma

	Hyperplasia	Lymphoma
Nodule location	Cortical	Cortical medullary and sometimes extranodal
Cellularity of nodules	Mixed	Uniform
Nodule density	Widely spaced	Close
Nodule contour	Irregular size and shape	Regular size and shape
Phagocytic debris	Usually present	Inconspicuous
Mantle zone	Present	Usually absent
Polarity	Present	Absent

cases of post-vaccinal lymphadenitis had been misdiagnosed as malignant lymphoma. Sinus histiocytosis with massive lymphadenopathy and angiofollicular lymph node hyperplasia can also give rise to mistaken diagnoses of malignant lymphoma.

The diagnosis of Hodgkin's disease rests, to a large extent, upon the identification of the characteristic Reed-Sternberg cells in the affected tissue. However, Reed-Sternberg cells or cells bearing a close resemblance to them may be occasionally observed in a variety of neoplastic and non-neoplastic conditions (Strum et al. 1970). This problem, among others, can lead to Hodgkin's disease being wrongly diagnosed (Symmers 1968).

Bone

The relative rarity of primary bone tumours means that few histopathologists have wide experience of these lesions, and yet the histological diagnosis is critical to the proper management of the patient. Spjut et al. (1971) emphasise the need for close co-operation between surgeon, radiotherapist and pathologist. Without exception, the X-ray appearances of the lesion must be examined by the pathologist before a diagnosis is made (Ackerman 1976).

Fracture callus has an infamous propensity for simulating osteosarcoma. Foci of proliferating osteoblasts closely admixed with osteoid can have an exceedingly sinister appearance and, without complete knowledge of the history and radiological features, can lead to the mistaken diagnosis of osteosarcoma. Additionally, callus can occur in association with pathological fractures, so that the underlying disease may be missed by the pathologist who is merely content not to have misinterpreted callus as a sarcoma.

The parosteal variant of osteosarcoma can present such a well-differentiated and mature appearance that it is likely to be dismissed as benign by those who are unaware of the entity. The lesion is not quite so indolent as its banal histological appearance might suggest.

Cartilagenous tumours are especially difficult. It is easy to underestimate the metastatic potential of a relatively acellular cartilagenous neoplasm in which mitotic activity and pleomorphism are inconspicuous. Conversely, chondromyxoid fibroma, a benign tumour, may be easily mistaken for chondrosarcoma (Jaffe and Lichtenstein 1948). Enchondromas of the hands and feet are common, whereas the malignant counterpart, chondrosarcoma, is relatively rare at these sites. The opposite is true in proximal sites, such as long bones and the axial skeleton, where chondrosarcomas are most common. It is therefore important to know the exact origin of the biopsy under consideration.

Other examples of potential diagnostic confusion are between osteoblastoma and osteosarcoma; Ewing's tumour and metastatic neuroblastoma, oat-cell carcinoma or "reticulum-cell" sarcoma; eosinophilic granuloma and chronic osteomyelitis; osteoclastoma and hyperparathyroidism, aneurysmal bone cyst, or giant-cell reparative granuloma of the jaw; ossifying fibroma and fibrous dysplasia; myositis ossificans and osteosarcoma (Spjut et al. 1971).

Breasts

Most of the diagnostic problems in breast pathology are complicated by the fact that they tend to be encountered during a rapid diagnosis by frozen section. Cryostat sections rarely have the clarity of paraffin sections from properly fixed tissues and it is therefore essential for the trainee histopathologist to see and compare as many cryostat sections as possible with their corresponding paraffin sections. Only in this way can one become familiar with the artifacts inherent in frozen sections.

One example of diagnostic difficulty, epitheliosis versus intraduct carcinoma, has been described earlier in this chapter. To assist in distinguishing between benign and malignant proliferation of ductal epithelium, the reader can do no better than refer to the excellent sequence of micrographs in the Armed Forces Institute of Pathology fascicle by McDivitt et al. (1968). The main points of distinction are summarised in Table 7.3, although, as McDivitt and his colleagues state, "Critical diagnosis, however, requires experience transcending mere tubular guidelines".

Table 7.3. Comparison of epitheliosis and intraduct carcinoma

	Epitheliosis	Intraduct carcinoma
Cell population	Mixed	Uniform
Nuclei	Normochromatic	Hyperchromatic
Pattern	Complex glandular	Regular cribriform or solid comedos
Microcalcification	Infrequent	More common
Cell streaming	Frequent	Absent

Distinguishing sclerosing adenosis from invasive carcinoma can be difficult, particularly with limited sampling and crytostat sections. The key to the problem lies in recognising the lobular distribution of the lesion within the biopsy tissue; this can be discerned only by low-power microscopy. Sclerosing adenosis typically has a lobular distribution so that, in a section of adequate size, multiple foci each separated by tracts of normal connective tissue will be evident (Fig. 7.11). Carcinoma, except for the lobular variant, will be a discrete solitary lesion in most instances.

Lobular carcinoma, as opposed to the more common ductal carcinoma, can be overlooked even in the invasive state. Carcinoma-in-situ of the lobules leads to distension of the acini and terminal ducts by a monomorphic cell population that is as

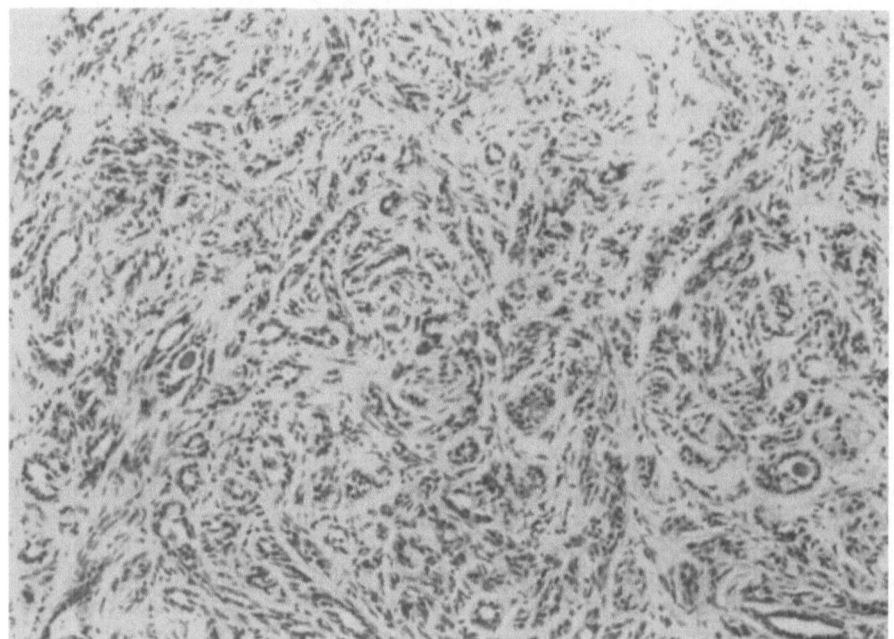

Fig. 7.11. Irregular glandular profiles in sclerosing adenosis mimic invasive mammary adenocarcinoma. The characteristic lobular disposition is usually visible at low magnification. Haematoxylin and eosin, ×85

banal as its counterpart in mammary ducts. Evidence of pagetoid spread along ducts is a helpful feature (Azzopardi 1979). Invasive lobular carcinoma typically consists of very slender files of infiltrating cells and, because very often no obvious tumour is visible either macroscopically or on low-power microscopy, the lesion may be completely missed or the appearances attributed to some inflammatory process (Fig. 7.12).

An intriguing phenomenon, fortunately rare, is the rather sinister finding of perineural invasion in benign breast lesions such as fibroadenosis or sclerosing adenosis. It is important for the pathologist to know that this is a possibility so that a diagnosis of malignancy is not made on this feature alone (Davies 1973). The glandular elements are found within the same perineural spaces that may be more commonly permeated by true malignant neoplasms. The biological significance of this in benign breast diseases is not known; it certainly appears to have no clinical significance.

Urinary Tract

The commonest primary malignant neoplasm of the renal parenchyma is the hypernephroma (renal adenocarcinoma). This tumour, characterised by clear cells

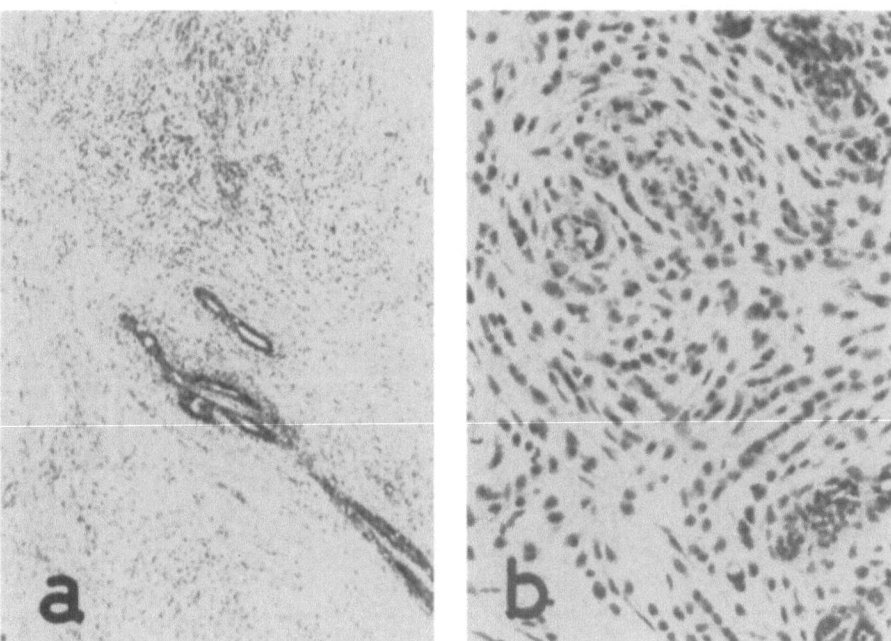

Fig. 7.12a,b. Invasive lobular mammary carcinoma. a Breast lump from a 45-year-old female originally reported as benign; 6 months later the patient returned with metastatic adenocarcinoma in the axillary lymph nodes. Haematoxylin and eosin, ×56. b Review of the breast lesion at higher magnification shows slender files of malignant cells, a pattern characteristic of invasive lobular carcinoma (see Fig. 5.10j). Haematoxylin and eosin, ×215

with a richly vascular stroma, commonly metastasises by way of the blood stream. Indeed, the secondary deposits may be the clinically presenting feature; the primary is often occult. When this is so, a clear-cell secondary deposit could be mistaken for a primary neoplasm of the particular organ in which it occurs. For example, sebaceous carcinoma, hidradenoma and "balloon-cell" melanoma in skin; clear-cell tumour and bronchioloalveolar-cell carcinoma in lung; acinic-cell tumour in salivary gland; haemangioblastoma in cerebellum; adenoma in adrenal; carotid-body tumour, glomus-jugulare tumour, parathyroid adenoma and medullary carcinoma of thyroid in the neck (Fig. 7.13). Some of these tumours can be identified by specific markers within them (e.g. melanosomes in melanoma, APUD granules and amyloid in medullary carcinoma). Other distinguishing features are well covered by Bennington and Beckwith (1975). Finally, it must be remembered that not all hypernephromas display the typical clear-cell features and a few may even grow in a spindle-cell fashion mimicking sarcoma (Fig. 7.3b).

Two inflammatory lesions of the kidney can resemble hypernephroma. Xanthogranulomatous pyelonephritis is characterised by a dense infiltrate of lipid-laden macrophages associated with a granulomatous inflammatory reaction, often with cholesterol crystal clefts. Malakoplakia is similar to xanthogranulomatous pyelonephritis in that foamy macrophages are prominent but, in addition, cytoplasmic inclusions (Michaelis-Gutmann bodies) are present. Both lesions, considered to be the result of obstruction and infection, may be erroneously interpreted as carcinoma.

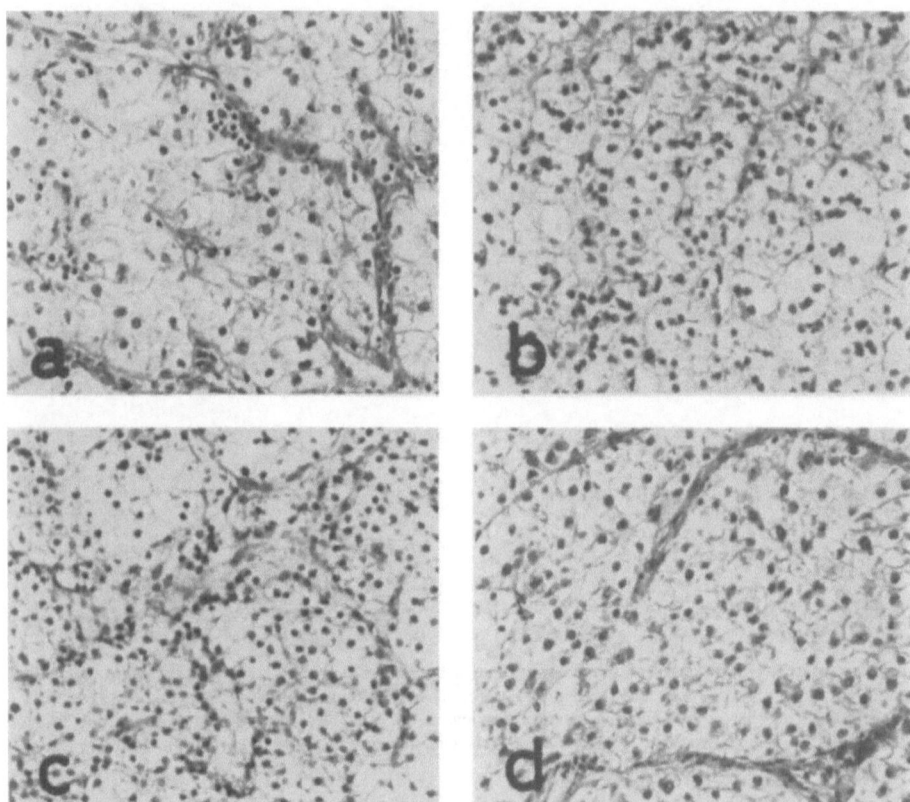

Fig. 7.13a–d. Mimicry in clear-cell lesions. a renal adenocarcinoma ("hypernephroma"); b misplaced adrenal cortex beneath renal capsule; c parathyroid adenoma; d "balloon-cell" malignant melanoma. Haematoxylin and eosin, ×215

The urothelium of the renal pelves, ureters and bladder is a common site of neoplasia; most of the lesions are transitional-cell carcinomas posing little diagnostic difficulty. Flask-shaped protrusions of urothelium into the lamina propria (Brunn's nests) may, in a small biopsy, be mistaken for superficially invasive transitional-cell carcinoma. However, though Brunn's nests are not neoplastic in themselves, carcinoma may arise within them.

Adenocarcinomas of the urinary tract can be simulated by glandular metaplasia of the urothelium, as seen in cystitis cystica or cystitis glandularis (Emmett and McDonald 1942).

Thyroid

Striated muscle may very closely invest the periphery of the thyroid gland so as to be mistaken for invasion of the muscle by a well-differentiated follicular carcinoma. Malignancy may also be misinterpreted from any small subcapsular deposits of thyroid tissue that may occasionally be found in cervical lymph nodes (Roth 1965);

these are considered to be a form of "benign metastasis" (Meissner and Warren 1969). Large deposits, especially if papillary structures or cytological features of malignancy are present, are much more likely to signal the presence of carcinoma within the thyroid gland.

An important point is the accurate recognition of follicular carcinoma of the thyroid and its distinction from a single adenoma. This can be extremely difficult. A key feature is invasion of blood vessels and capsule, though the latter is held by some to have less significance. Elastic stains are useful to delineate the walls of blood vessels within or around a thyroid tumour, particularly when the lumen has been occluded by a plug of neoplastic follicular tissue. Capsular invasion can sometimes be difficult to distinguish from passive entrapment of glands in the capsule as part of a regressive process.

Male Genitalia

The pseudomalignant appearance of malakoplakia has been discussed. A related entity, granulomatous orchitis, is characterised by testicular infiltration by plasma cells, lymphocytes and macrophages. Hodgkin's disease may be misdiagnosed, or the large vacuolated epithelioid macrophages may be mistaken for the cells of a Sertoli tumour (Mostofi and Price 1973).

Squamous metaplasia in the prostate, often found close to areas of infarction, can lead to a mistaken diagnosis of invasive carcinoma. It may also be seen in prostates from patients treated with oestrogens. Difficulty may also be experienced in distinguishing atypical or distorted glands in a hyperplastic prostate from a well-differentiated adenocarcinoma. However, in a hyperplastic gland the acini are invariably lined by two cell layers (epithelium and myoepithelium) whereas in carcinoma only a single layer is present (neoplastic epithelium) (Fig. 7.14); the epithelial cells in an adenocarcinoma thus abut directly onto the stroma. Acini in striated muscle must not always be accepted as evidence of invasive carcinoma because striated muscle fibres normally occur at the external boundary of the prostate.

Female Genital Tract

Changes in the endocrine milieu, either natural or induced by therapy, are important causes of pseudomalignant changes in the female genital tract. For example, a decidual reaction is easy to recognise in the endometrium but this may not be so in unfamiliar locations; in the cervix decidual reactions can be wrongly interpreted as neoplastic infiltration (Fig. 7.15; Lapan 1949).

The Arias-Stella reaction of the endometrium, seen in association with oral and intrauterine contraceptives and with pregnancy, can present a disturbing histological picture. Nuclear enlargement, polyploidy, and hyperchromaticism in the context of an exuberant glandular proliferation may falsely invite a diagnosis of malignancy. This benign proliferative reaction tends to be characterised by pallor and "ferning" of the glandular epithelium, features which, while not absolutely diagnostic, should arouse one to the possibility of an Arias-Stella reaction and solicit a more cautious interpretation of the nuclear abnormalities.

At the end of pregnancy or subsequent to abortion, placental remnants are sometimes seen in uterine curettings. The naturally invasive trophoblast may penetrate

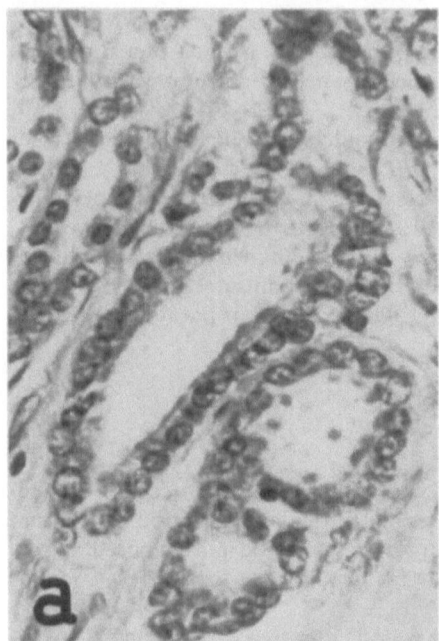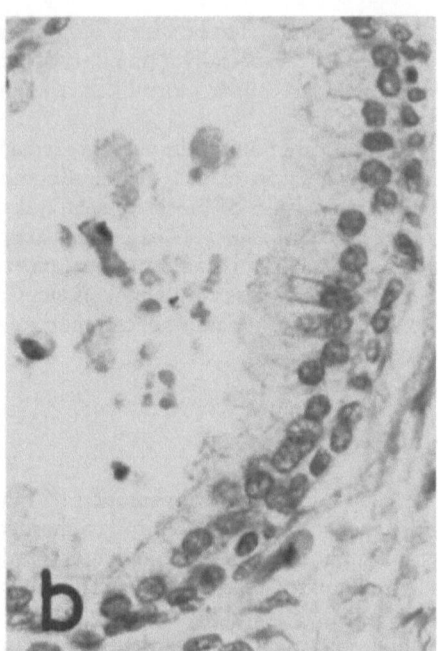

Fig. 7.14a,b. Adenocarcinoma of the prostate versus senile hyperplasia. **a** Well-differentiated adeno-carcinoma in which the glands are characteristically lined by a single cell layer. **b** Hyperplastic glands which, like those of the normal prostate, are endowed with a double cell layer. Haematoxylin and eosin, ×558

the superficial myometrium, a placental site reaction that may simulate chorio-carcinoma (Elston 1978). Distinction between placental site reactions and chorio-carcinoma requires considerable experience of the normal range of appearances of placental implantation sites.

Large doses of progestogens sometimes result in a peculiar sarcoma-like change in the endometrial stroma. The bizarre stromal cells, in the absence of a history of progestogen therapy, would certainly be regarded with considerable suspicion (Charles 1964).

Two benign proliferative changes in uterine cervical glands can be mistaken for carcinoma. First, the glands can become filled by proliferation of the outer layer of reserve cells; the solid cell nests that result may mimic the histological picture of invasive carcinoma. Secondly, a florid reactive proliferation of small mucous glands that can be mistaken for adenocarcinoma is occasionally seen, for example, associated with oral contraceptives or pregnancy (Nichols and Fidler 1971).

Ectopic endometrium in unfamiliar locations (i.e. endometriosis) may falsely incite suspicion of malignancy, especially adenocarcinoma. Endometriotic foci, however, are usually devoid of the cytological features of malignancy and the glands are typi-cally invested by endometrial stroma, often with haemosiderin nearby.

Urethral caruncles can simulate squamous-cell carcinoma. Deep invaginations of the surface epithelium give an appearance like that of neoplastic invasion, particularly if some show no obvious connection with the surface owing to an unfavourable plane of section.

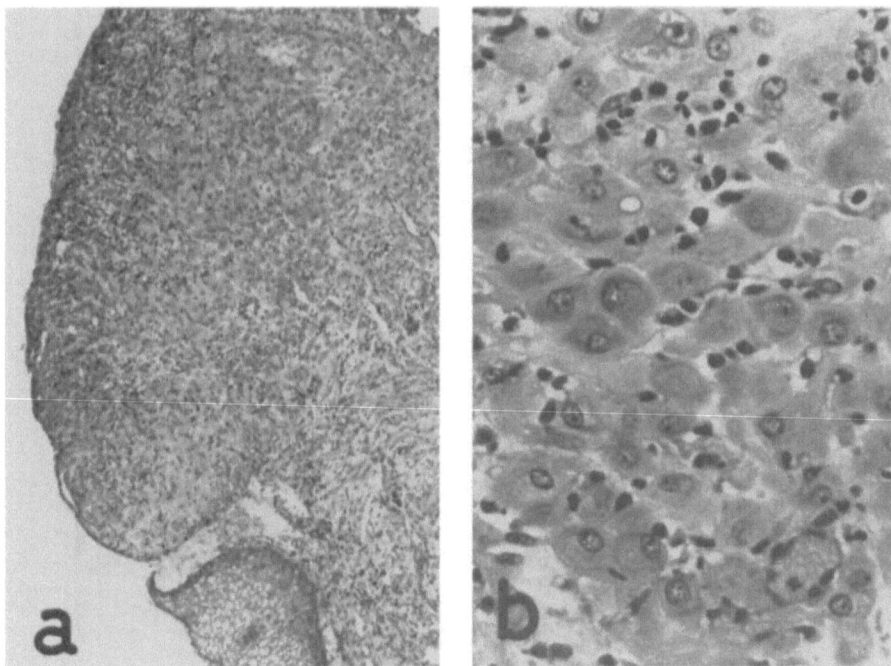

Fig. 7.15a,b. Deciduosis of the uterine cervix in pregnancy. a Cervical biopsy showing a decidual nodule breaking through the epithelial surface. Haematoxylin and eosin, × 34. b Decidual cells giving an appearance that might be mistaken for a malignant tumour. Haematoxylin and eosin, × 215

Adenocarcinoma of the fallopian tube, a rare lesion, may be simulated by chronic inflammation causing fusion of the plicae, creating a cribriform pattern with reactive epithelial hyperplasia.

Gastrointestinal Tract

Primary carcinomas of the oesophagus are readily identified in biopsies. However, the histopathologist should be conscious of the possibility that a carcinoma clinically presenting as an oesophageal lesion may, in fact, be a direct extension from a mediastinal lymph node involved by metastatic carcinoma from the lung with erosion of the adjacent oesophageal wall. Clear evidence of an origin of the carcinoma from the overlying or adjacent oesophageal epithelium is desirable if doubt as to the origin of the neoplasm is to be eliminated.

Endoscopic biopsies of gastric mucosa are now common and contribute much to the management of ulcerated lesions of the stomach. Assessment of these small biopsies is often difficult owing to the frequent occurrence of glandular distortion and dysplasia in the regenerating mucosa adjacent to benign gastric ulcers. A useful way of gaining experience in the interpretation of these biopsies is to familiarise oneself with the mucosal appearances commonly present adjacent to unequivocally benign ulcers in well-fixed gastrectomy specimens.

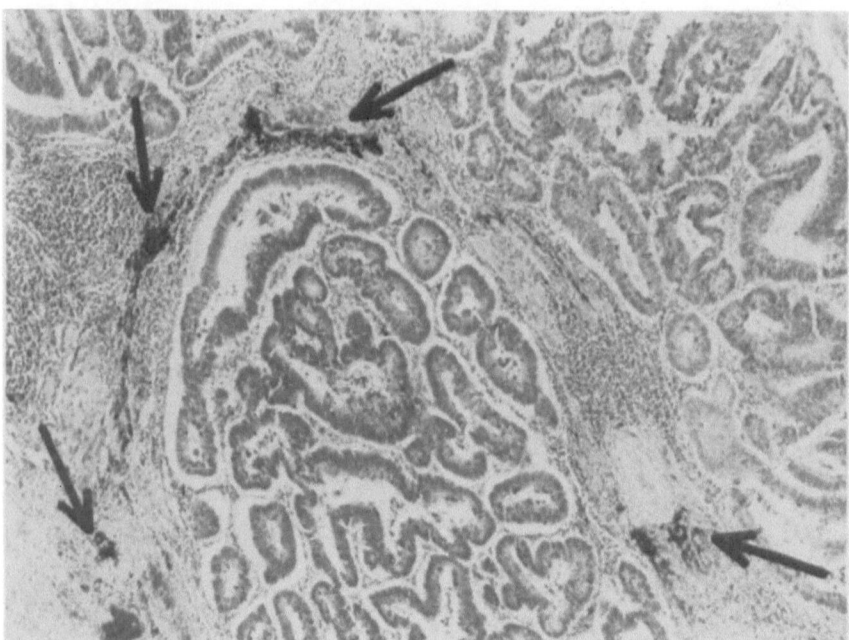

Fig. 7.16. Pseudo-invasion in an adenomatous rectal polyp. A group of glands, surrounded by deposits of haemosiderin (*arrowed*), has become detached from the surface adenomatous tissue probably due to previous torsion. Perls' stain, ×65

Distorted glands lined by cells with hyperchromatic pleomorphic nuclei may be seen in or adjacent to ulcerated gastric mucosa and mistaken for adenocarcinoma in small biopsies. Atypical cells, possibly endothelial, may also be seen in endoscopic biopsies of ulcerated mucosa of the oesophagus, stomach and rectum and may be misdiagnosed by the unwary as malignant (Isaacson 1982).

The identification of malignancy in adenomatous polyps of the large bowel has an important bearing on the treatment of the patient. The only unequivocal evidence of malignancy is invasion of the stalk or submucosa. Neoplastic invasion in polyps can be closely simulated by the passive entrapment of glands within the stalk of a lesion that has been damaged by torsion or other trauma (Muto et al. 1973). Haemosiderin can often be demonstrated in the vicinity of such glands and this helps to differentiate the lesion from true carcinomatous invasion (Fig. 7.16). Glandular distension by inspissated mucin can simulate the appearance of a "colloid" carcinoma.

Respiratory Tract

The majority of nasal polyps are benign inflammatory or allergic lesions. However, polyps in which the core is traversed by numerous broad trabeculae of transitional-type epithelium, i.e. inverted (Ringertz) papilloma, are occasionally seen. This epithelial inversion may be mistaken for malignant neoplastic invasion. However, although this lesion does have a tendency to recur locally it does not metastasise.

Carcinoid tumours arising in bronchi can, when seen in a crushed bronchial biopsy, be difficult to distinguish from oat-cell carcinoma, or, less frequently, adenocarcinoma. Although these tumours can be aggressive and metastasise, distinction from other pulmonary neoplasms is obviously important, particularly from the point of view of management. Small aggregates of cells resembling oat cells or carcinoid tumours may be found in peripheral lung tissue. These generally innocent lesions, often referred to as "tumourlets", have an uncertain histogenesis. Bronchiectasis seems to be a common predisposing lesion.

Serosal Surfaces

Simple mesothelial proliferation, such as occurs in response to inflammation, may give rise to patches of multilayering that resemble neoplastic nodules, particularly if they exhibit a pseudoglandular appearance. The absence of any of the cytological hallmarks of malignancy is helpful in recognising the true nature of the lesion.

Organisation of inflammatory exudate on mesothelial surfaces can lead to the entrapment of groups of mesothelial cells and mimicry of neoplastic infiltration. This is most often seen in hydrocele sacs. Critical scrutiny usually reveals that the cells are confined to a single level in the sac wall, presumably the original location of the old serosal surface (Fig. 7.17).

Mesothelial cells in serous effusions have a notorious facility for simulating malignant cells (Chap. 6). Cytoplasmic vacuolation may be sufficiently pronounced to give rise to a "signet ring" appearance; periodic acid-Schiff staining cannot be reliably used to distinguish these cells from a mucus-secreting adenocarcinoma because both can give positive results. Mucicarmine is better. Mitotic activity, clumping, and multinucleate cells may also be evident in non-neoplastic effusions, but with experience these changes can be recognised as reactive; absence of nuclear hyper-

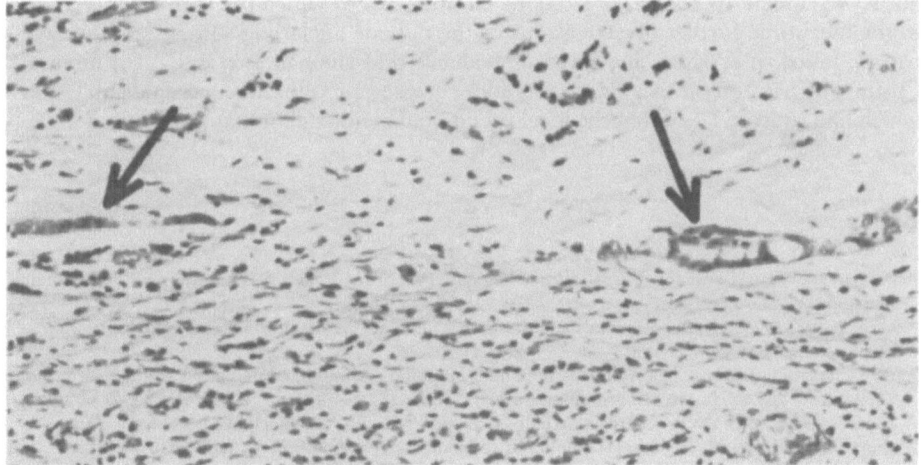

Fig. 7.17. Sequestration of mesothelial cells (*arrowed*) in the thickened wall of an hydrocele sac. This probably results from organisation of exudate on the original mesothelial surface. Haematoxylin and eosin, ×215

chromaticism, an even dispersion of nuclear chromatin, and a narrow perinuclear halo are characteristics of mesothelial cells (Koss 1968).

Central Nervous System

Neuropathology is a highly specialised field and so, for the novice in general histopathology, it will suffice to allude to a single example of pseudomalignancy within the central nervous system. The glial component of the central nervous system is the commonest origin of primary neoplasms within the brain and it is also the component that most actively participates in the response of the brain to local damage. Thus, reactive gliosis and astrocytoma are occasionally difficult to separate, particularly if the latter is a relatively acellular lesion. The detailed cytology of the astrocytes is of considerable importance, for reactive gliosis invariably lacks the mitotic activity, pleomorphism and hyperchromaticism that would favour a neoplasm.

Conclusions

It must be emphasised that this account of pseudomalignancy and related problems is only a guide around the principal pitfalls. Pattern recognition may be an easy method of making many diagnoses at the microscope; however, the pattern of benign and malignant lesions may be confusingly similar, only to be resolved by critical and careful analysis of the histological appearances. Additional clinical information, sections, or tissue often clarify an otherwise insoluble differential histological diagnosis. Ackerman and Tanski (1977) related the salutory tale of a 37-year-old woman, a doctor's wife (inevitably), in whom an incorrect diagnosis of a malignant lymphoproliferative process was made on a skin biopsy. However, this erroneous diagnosis was upheld by several prestigious laboratories to which the case was referred. In the meantime further investigations of the patient, including some of an invasive nature, failed to produce one shred of evidence that she had any malignant disease. Cutting further into the original tissue block revealed a molluscum contagiosum lesion which had ruptured, exciting the local pseudoleukaemic inflammatory reaction!

References

Ackerman AB, Tanski EV (1977) Pseudoleukaemia cutis: report of a case in association with molluscum contagiosum. Cancer 40: 813–817

Ackerman LV (1976) Common errors made by pathologists in the diagnosis of bone tumours. Recent Results Cancer Res 54: 120–138

Allen AC (1948) Persistent "insect bites" (dermal eosinophilic granulomas) simulating lymphoblastomas, histiocytoses and squamous cell carcinomas. Am J Pathol 24: 367–387

Azzopardi JG (1979) Problems in breast pathology. Saunders, London Philadelphia Toronto

Bennington JL, Beckwith JB (1975) Tumours of the kidney pelvis and ureter. Armed Forces Institute of Pathology, Washington DC, pp 178–186

Charles D (1964) Iatrogenic endometrial patterns. J Clin Pathol 17: 205–212

Cocker J, Fox H, Langley FA (1968) Consistency in the histological diagnosis of epithelial abnormalities of the cervix uteri. J Clin Pathol 21: 67–70

Davies JD (1973) Neural invasion in benign mammary dysplasia. J Pathol 109: 225–231

Davies SW (1965) Giant condyloma acuminata: incidence among cases diagnosed as carcinoma of the penis. J Clin Pathol 18: 142–149

Dorfman RF, Warnke R (1974) Lymphadenopathy simulating the malignant lymphomas. Hum Pathol 5: 519–550

Elston CW (1978) Gestational tumours of trophoblast. In: Anthony PP, MacSween RNM (eds) Recent advances in histopathology, No. 11 Churchill Livingstone, Edinburgh London Melbourne New York, pp 149–161

Emmett JL, McDonald JR (1942) Proliferation of glands of the urinary bladder simulating malignant neoplasm. J Urol 48: 257–265

Enzinger FM (1970) Epithelioid sarcoma: a sarcoma simulating a granuloma or a carcinoma. Cancer 26: 1029–1041

Hart WR, Norris HJ (1973) Borderline and malignant mucinous tumours of the ovary. Cancer 31: 1031–1045

Hartsock RJ (1968) Postvaccinial lymphadenitis: hyperplasia of lymphoid tissue that simulates malignant lymphomas. Cancer 21: 632–649

Hyams VJ (1975) Spindle cell carcinoma of the larynx. Can J Otolaryngol 4: 307–313

Isaacson PG (1982) Biopsy appearances easily mistaken for malignancy in gastrointestinal endoscopy. Histopathology 6: 377–389

Jaffe HL, Lichtenstein L (1948) Chondromyxoid fibroma of bone. A distinctive benign tumour likely to be mistaken for chondrosarcoma. Arch Pathol 45: 541–551

Koss LG (1968) Diagnostic cytology, 2nd edn. Lippincott, Philadelphia, pp 497–499

Langley FA (1978) Quality control in histopathology and diagnostic cytology. Histopathology 2: 3–18

Lapan B (1949) Deciduosis of the cervix and vagina simulating carcinoma. Am J Obstet Gynecol 58: 743–747

McDivitt RW, Stewart FW, Berg JW (1968) Tumours of the breast. Armed Forces Institute of Pathology, Washington DC, pp 22–47

McGovern VJ (1976) Malignant melanoma: clinical and histological diagnosis. Wiley, New York London Syndey Toronto, pp 147–150

Meissner WA, Warren S (1969) Tumours of the thyroid gland. Armed Forces Institute of Pathology, Washington DC, pp 25–27

Morson DB, Dawson IMP (1979) Gastrointestinal pathology, 2nd edn. Blackwell, Oxford London Edinburgh Melbourne, pp 212–215

Mostofi FK, Price EB (1973) Tumours of the male genital system. Armed Forces Institute of Pathology, Washington DC, pp 13/–139

Muto T, Bussey HJR, Morson BC (1973) Pseudo-carcinomatous invasion in adenomatous polyps of colon and rectum. J Clin Pathol 26: 25–31

Nichols TM, Fidler HK (1971) Microglandular hyperplasia in cervical cone biopsies taken for suspicious and positive cytology. Am J Clin Pathol 56: 424–429

Park WW (1980) The histology of borderline cancer. Springer, Berlin Heidelberg New York

Roth LM (1965) Inclusions of non-neoplastic thyroid tissue within cervical lymph nodes. Cancer 18: 105–111

Saltzstein SL, Ackerman LV (1959) Lymphadenopathy induced by anticonvulsant drugs and mimicking clinically and pathologically malignant lymphomas. Cancer 12: 164–182

Spjut HJ, Dorfman HD, Fechner RE, Ackerman LV (1971) Tumours of bone and cartilage. Armed Forces Institute of Pathology, Washington DC

Stout AP, Lattes R (1967) Tumours of the soft tissues. Armed Forces Institute of Pathology, Washington DC, pp 146–147

Strum SB, Park JK, Rappaport H (1970) Observations of cells resembling Sternberg-Reed cells in conditions other than Hodgkin's disease. Cancer 26: 176–190

Symmers WStC (1968) Survey of the eventual diagnosis in 600 cases referred for a second histological opinion after an initial diagnosis of Hodgkin's disease. J Clin Pathol 21: 650–653

8   Rapid Frozen Section Diagnosis

Three factors contributed to the emergence of frozen sections as a rapid diagnostic method in the final decades of the nineteenth century. First, anaesthesia and aseptic techniques enabled surgeons to perform longer and more ambitious operations. Second, improvements in the optics of microscopes due to innovations such as the Abbé condenser and achromatic lenses made it possible for histologists to decipher the images of relatively inferior sections. Third, tissue preparation techniques had been considerably refined. Thus it became possible for surgeons to exploit the full diagnostic potential of histopathology actually during the course of surgical operations. Before about 1870, histological sections were cut by hand using razor blades; then the microtome appeared in histology laboratories, enabling better quality sections to be cut more rapidly; and with facilities for freezing and maintaining the tissues in a frozen state while being cut, it was possible to cut good sections almost immediately the tissue had been received in the laboratory (Wright 1985). The sections could be fixed rapidly in formalin (first reported as a fixative in 1893) before staining.

Although William H. Welch is credited with the first application of the frozen section technique for rapid intraoperative diagnosis (the patient was being operated on by William S. Halstead for a breast tumour—it was reportedly benign but the operation was completed before Welch's "rapid" diagnosis reached the operating theatre!), it was Thomas Cullen, who studied in Welch's laboratory, who promulgated the technique in the English (American) and German literature (Cullen 1895 a & b). In Germany itself the technique was being developed by Henrique Plenge in Heidelberg and Ludwig Pick in Berlin (Pick 1897). In Great Britain, Alfredo Kanthack from St. Bartholomew's Hospital in London and T. Strangeways Pigg published their rapid method in 1897 (Kanthack and Pigg 1897); the tissue was immersed in boiling water for a few minutes and then frozen sections were cut on a freezing microtome. But it was Ernest Shaw, pathologist at the Great Northern Hospital in London, who in

1899 first used rapid frozen sections to considerable clinical advantage while working with Charles Lockwood, a general surgeon (Shaw 1910), and by 1923 he was able to publish his experience of 540 cases (Shaw 1923).

My first case, undertaken at Yeovil in 1899 for Mr. Lockwood, will serve as a good example. A tumour was present in the breast of a lady; its physical characteristics were those common to a mass of carcinoma or chronic mastitis. It was explored and a piece removed for microscopic examination. This proved to be a carcinoma. Fortified by this knowledge the surgeon was then able to proceed with the formidable operation of amputation of the breast with removal of the pectoral muscles and axillary glands, satisfied in his own mind that he was doing the correct operation for his patient (Shaw 1910).

Shaw vividly conveyed the novelty of the technique and the difficult conditions under which he and his contemporaries did frozen sections.

Quite a small table will do to arrange the various instruments on, and this should be placed opposite a window if possible. I have had to fix my apparatus in all kinds of odd places, such as the landing at the top of a flight of stairs, a board placed across an ordinary bath, the corner of an operating room, a rickety table in a small cottage, on a bed, etc. Most often one can commandeer a bedroom after the patient has been taken out to the operating room. Sunlight from a window is the best, but sometimes one has to trap a beam from an overhead lamp. A paraffin lamp in a house far away in the country caused me the greatest trouble. I have recently seen a new form of illumination in which an electric bulb is fitted to the substage reflector of the microscope, the current is supplied by a small portable battery. This ought to be very convenient. All my apparatus, including a folding microscope, is carried in a small bag (Shaw 1923).

In North America the value of rapid frozen sections was hotly contested; some surgeons and pathologists denounced the technique as a mere fad, while others—notably the staff of the Mayo Clinic—strongly endorsed it (Mayo 1929; Dahlin 1980). By 1905 the technique had become established at the Mayo Clinic (Jennings and Landers 1957) and three years later, at the Presbyterian Hospital New York (Nakazawa et al. 1968).

Competing for a time with frozen sections as a rapid diagnostic method was the razor-section technique devised by Benjamin Terry, professor of pathology at Vanderbilt University, U.S.A. (Terry 1927). Terry's technique involved cutting thin tissue slices with a razor blade and then applying a polychrome methylene blue stain to one surface of the slices. When transilluminated by a bright light and viewed microscopically the cells exposed and stained on the cut surface were clearly seen as though in a thin section. This was a very rapid method requiring much less equipment and manual dexterity than for the preparation of frozen sections. Although the stained preparations were not permanent the tissue slice could be then processed routinely and permanent sections cut for diagnostic confirmation and archival purposes. Also, the technique was unsatisfactory with very vascular or pigmented tissue which could not be transilluminated adequately. Popular in Europe was the Leitz Ultropak illuminator which permitted the microscopic examination of opaque objects; thus relatively thick pieces of tissue could be used because the need for transillumination was eliminated (Wright 1985). A similar technique was published by Mac-Mahon and DelVecho (1944) from the Cambridge Hospital in Massachusetts, U.S.A., and was still used into the 1950s in some hospitals in New England and eastern Canada (Wright 1985). However, the overall supremacy of the frozen section technique was probably attributable to two factors: the other methods of rapid diagnosis did not directly yield permanent preparations and the microscopic examination of tissue surfaces was perhaps a little too unconventional for many practitioners who had been reared on histological sections.

The basic methods used by Shaw and others were used with only minor modifications until the late 1950s when the cryostat (refrigerated microtome) became commercially available.

Clinical Indications

The only acceptable clinical indication for doing a rapid frozen section is the need for a biopsy opinion which will result in an immediate therapeutic decision, usually surgical. The specific indications can be broadly categorised.

1. *To determine the nature of a disease.* This is the commonest indication; it comprises the bulk of frozen section work in most laboratories. The most frequent application is still in many places the intra-operative diagnosis of breast lesions, so that benign lesions can be dealt with by nothing more than local excision and malignant tumours treated more aggressively and without delay.

2. *To identify a tissue.* Intra-operative confirmation of the nature of an excised piece of tissue is probably the second commonest indication. For example, frozen section facilities are virtually essential in parathyroid surgery; brown fat, peripheral thyroid nodules, lymph nodes, and ectopic thymus can all macroscopically resemble pathological parathyroid tissue. Autonomic nervous tissue can be rapidly confirmed by frozen section for patients undergoing vagotomy or sympathectomy.

3. *To determine the extent of a disease.* Identification of ganglia in bowel wall during resections for Hirschsprung's disease determines the limits of an aganglionic segment. In cancer surgery, assessment of tissue at resection margins and lymph nodes may help to guide the surgeon performing extirpative operations.

4. *To confirm that lesional tissue has been removed.* If the sole purpose of a surgical operation is to remove tissue for a diagnostic biopsy, the procedure is wasted if the specimen ultimately proves to be inadequate. Although a frozen section is not often done for this reason, it is a useful way of checking that a satisfactory sample has been obtained before the anaesthetic is terminated. For example, a lymph node can be identified as abnormal on a frozen section without going so far as to identify the particular disease process. Biopsy tissue can also be rapidly assessed for viability; necrotic material is generally useless for diagnostic purposes.

The overall clinical value of rapid frozen section diagnosis is undisputed; it reduces the need for and risks of a second anaesthetic, and operative morbidity and mortality is reduced. Intra-operative diagnosis of suspected malignant neoplasms lessens the patient's emotional attrition while waiting for the biopsy report that may subsequently dictate the need for a mutilating cancer operation. Hospital stay is shortened, an economic benefit that more than offsets the cost of maintaining an efficient frozen section service.

Doing frozen sections out of idle curiosity, to fill in a coffee break, to increase a surgeon's, pathologist's, or hospital's income, or merely to prove that the pathologist is actually in the hospital, are spurious indications (Saltzstein and Nahum 1973).

Methods

The early methods used sections cut by hand from frozen tissue and stained with polychrome dyes. The Linderstrøm-Lang cold-chamber cryostat was devised in 1938 to meet the needs of histochemistry, but it was not until the 1960s that reports of the use of cryostats in diagnostic histopathology first appeared (Chang et al. 1961). The cryostat became so popular that in some centres, notably at the University of Texas M. D. Anderson Hospital, where the machine was first used in clinical work (Sparkman 1962), frozen sections replaced paraffin sections even for routine surgical pathology.

Having received a specimen with a request for a rapid frozen section diagnosis, the pathologist should not feel under any absolute obligation to cut a section in every instance. A frozen section request is, in essence, the surgeon's way of asking the pathologist for an opinion of the nature of the lesion. Whether or not a frozen section is actually cut should be left entirely to the pathologist's discretion. For instance, a breast lump which is just a circumscribed homogeneous fatty nodule is almost certainly a lipoma. Any attempt to do a frozen section will be thwarted by the inherent difficulties of cutting fatty tissue in a cryostat. If a section will actually stay on the slide while it is being stained, the end result is unlikely to be of diagnostic quality.

Sampling of the fresh tissue is governed by the same principles that govern the selection of blocks from fixed tissue for routine paraffin sections. The gross specimen must be carefully inspected and, if sufficiently large, incised. Palpation is useful, especially for breast lesions, but the most easily palpable or visible lesion may distract from the most clinically significant. Minute carcinomas can lurk amidst fibroadenosis. Lobular carcinomas of the breast are sometimes macroscopically banal, with no obvious localised tumour.

Palpate the tissue gently; rough handling of unfixed samples will produce crushing artifacts. Always cut the tissue with a clean sharp blade; a blunt knife will distort the tissue. Endeavour to save at least a small portion of the lesion for paraffin sections to avoid freezing artifacts, and always process the piece that has been frozen as a check on sampling.

Even with a cryostat, some tissues remain difficult to cut. Bone obviously poses problems, but small spicules are not an insurmountable problem. Skin can be difficult, but most experienced technicians can produce satisfactory sections. Sections of fatty or necrotic tissues tend to fragment; lesions should be trimmed of fat and necrosis before sectioning.

Unfixed tissue is always preferable to fixed material for cryostat work, but this can expose staff to certain microbiological hazards and possible contamination of the equipment (Duray, Flannery and Brown 1981). Tuberculous lesions, for example, should be avoided; if the diagnosis is certain, a frozen section is obviously unnecessary. Sometimes, however, the diagnosis of tuberculosis is made unexpectedly and the contaminated cryostat should be appropriately disinfected before being used again. Hepatitis virus B is a very real danger; gloves must always be worn when handling fresh tissue and aerosols must be minimised by carefully opening bags or pots containing specimens. Frozen sections on urgent diagnostic problems in known carriers of hepatitis virus B should be done only after immersing the tissue in boiling formalin, a traditional method of rapid fixation.

Interpretation

Ackerman and Ramirez (1959) rightly emphasise that frozen sections should never be delegated to trainees, but should always be reported by senior histopathologists "rich in experience, conservative in attitude and, most important, he must have judgement". Trainees must therefore gain their experience and tuition by working at the elbow of their seniors, not by trial and error on their own.

Adequate clinical data must always be made available when frozen sections are reported; it is just as essential as it is for the more leisurely reporting of paraffin sections. Surgeon and pathologist should confer before embarking on a diagnostic frozen section in an unusual case. Any previous biopsy material, particularly if neoplastic, should be reviewed beforehand and kept available for reference at the time of doing the frozen section.

Generally, cells tend to appear larger in frozen sections than they do in paraffin sections of the same tissue and this may produce an inherent bias towards a mistaken malignant interpretation of benign lesions. It is advisable to compensate mentally for this and, if errors are to be made, always to err on the side of a benign diagnosis. A single false-positive diagnosis of cancer may be sufficient to eradicate a surgeon's confidence in the pathologist.

Not infrequent interpretative errors involve various well-differentiated malignant tumours that are called benign lesions by the overcautious pathologist. Alternatively, very poorly differentiated tumours, especially if admixed with a dense inflammatory infiltrate, can be dismissed as non-neoplastic lesions. Inflammatory reactions alone can also mimic cancer; plump endothelial cells in granulation tissue sometimes bear a remarkable resemblance to columns of infiltrating neoplastic cells.

Only two types of report are acceptable for a rapid frozen section diagnosis on a suspected malignant neoplasm.

1. A *definite diagnosis* on which the surgeon can take therapeutic action.
2. Advice to *wait for paraffin sections*.

Vacillation, equivocation, and indecision are to be avoided. "Possibly malignant" is a dangerous statement and could precipitate an aggressive surgeon into radical surgery that might well prove to be unnecessary. If a diagnosis cannot be made after 10 or 15 minutes' perusal of the section, further study is unlikely to enable the pathologist to render an opinion that is sufficiently reliable for therapeutic purposes. The difficulty may lie either in the quality of the biopsy material or in the intrinsic nature of the lesion. In the former instance, further samples can be taken for frozen section. In the latter instance, the pathologist and surgeon should ensure that enough biopsy tissue has been removed for paraffin sections, electron microscopy, etc. in what is probably going to be a difficult diagnostic exercise.

Although speed is important it must not be allowed to prejudice the accuracy of the given diagnosis. The pathologist's judgement should not be adversely influenced by a sense of urgency; only calm interpretation of the material will allow a thoroughly considered diagnosis. Often it is possible to give a quick answer to the question "can a *definite* diagnosis be made on this material?" before proceeding further, thus eliminating unnecessary delay and equivocation.

Common Applications

Breast Lesions

While pre-operative needle biopsies have tended to reduce the need for intra-operative frozen sections, breast lesions still constitute the largest single group in most laboratories and provide some of the most taxing problems. Azzopardi (1979) rightly asserts that most errors occur through either reporting on a cryostat section without seeing the gross specimen or not taking fully into account the low-power microscopy of the lesion. When errors occur in the interpretation of frozen sections of breast lesions they tend to result in the overdiagnosis of cancer, in contrast to the trend towards underdiagnosis with paraffin sections.

Sclerosing adenosis, the perennial histological mimic of invasive carcinoma, rarely has the gross appearance of a malignant neoplasm; low-power scanning usually reveals the lobular distribution of the lesion. Epithelial proliferation should always be interpreted bearing in mind the context in which it is seen; it is commonly a marked feature of fibroadenomas or gynaecomastia and should not be considered unduly sinister in these circumstances.

Thyroid

Follicular carcinoma can rarely be distinguished from an adenoma or adenomatous nodule unless vascular invasion is seen in a favourable section. Involvement of tissue outside the gland, lymph nodes for example, favours a malignant process. Papillary and anaplastic carcinomas are easier to recognise as malignant tumours on frozen sections.

Small-cell anaplastic carcinoma, lymphoma and Hashimoto's thyroiditis constitute a common problematic diagnostic triad. Small-cell anaplastic carcinoma and lymphoma are often difficult to differentiate, even in paraffin sections, without resorting to immunohistochemistry or electron microscopy. Anyhow, there is less of an urgent need to classify the malignancy on cryostat sections precisely. More important for the immediate treatment of the patient is the distinction between Hashimoto's thyroiditis and either of the two malignancies already cited. Favouring Hashimoto's disease are lymphoid follicles with reactive germinal centres, plasma cells, Hürthle-cell metaplasia and symmetrical enlargement of the gland.

Parathyroid Glands

Neck explorations in patients with hyperparathyroidism often call for the assistance of rapid frozen sections, first to identify a piece of tissue as parathyroid, and secondly to decide whether the gland is normal, hyperplastic, or an adenoma.

Wang and Rieder (1978) claim that the density of the gland can be a useful guide to the presence of hyperplasia or an adenoma. Normal prathyroid glands float in a mannitol solution of specific gravity 1.049–1.069 whereas abnormal parathyroids will sink; this is because of the relative paucity of stromal fat in hyperplasia and adenomas. Of course, one must be certain that parathyroid tissue is being assessed and not simply a nodule of fat that has been mistaken for a parathyroid gland.

Clear-cell hyperplasia poses relatively few problems because it invariably produces a diffuse enlargement of the glands. However, chief-cell hyperplasia produces a nodular adenomatous configuration virtually indistinguishable from an adenoma (Castleman and Roth 1978). Adenomas can be distinguished from chief-cell hyperplasia with certainty only by finding a second gland that is either normal or suppressed. A second enlarged gland strongly favours hyperplasia rather than adenoma.

Of limited aid is the demonstration of intracytoplasmic fat droplets in chief cells (King and Hirose 1979). The suppressed chief cells of a patient with an adenoma often contain an excess of intracytoplasmic fat. However, variable quantities of intra-cytoplasmic fat can be demonstrated in adenomas and adenomatous glands, so the technique may not be absolutely reliable. It may well be a useful way of identifying a compressed rim of suppressed normal parathyroid tissue around an adenoma.

Some adenomas have an acinar arrangement of cells, closely mimicking thyroid even to the extent of containing colloid material. This can be confusing.

Parathyroid glands should be measured and weighed before tissue is removed for cryostat sections. The dimensions of a normal gland do not usually exceed $6 \times 4 \times 2$ mm. Each parathyroid gland normally weighs about 30 mg, the combined weight of all the glands from a normal adult male being 120 ± 3.5 mg (mean $\pm$ SEM.) and from a female 142 ± 5.2 mg; the amount of adipose tissue in each gland increases with age.

Gastrointestinal Tract

Frozen sections may be requested on gastric ulcers when the distinction between a chronic peptic ulcer and ulcerated carcinoma is in doubt. As with most of these problems the distinction is easiest if tumour can be demonstrated in a regional lymph node or peritoneal deposit; this also reduces the risk of implanting tumour by incising the primary neoplasm. Note especially that "signet ring" or diffusely infiltrating desmoplastic carcinomas consist of tumour cells that may be almost imperceptible in frozen sections.

Biopsies from the pancreas and ampullary region must rank as the bane of cryostat diagnosis. Acinar and duct distortion in inflammatory conditions presents an extremely disconcerting appearance, only to be matched by the tendency of some desmoplastic well-differentiated adenocarcinomas to simulate chronic pancreatitis (Spjut and Ramos 1957). Accessory ducts, normally present, can also be mistaken for invasive carcinoma (Loquvam and Russell 1950). This is indeed treacherous territory for both surgeon and pathologist. The diagnosis of malignancy must be based only on absolutely unequivocal appearances; there is no room for speculation. Remember that the surgical treatment for a *diagnosis* of malignant neoplasm in this region is pancreaticoduodenectomy, and variations on that theme, an operation not without significant inherent morbidity and mortality.

A common procedure in paediatric practice is the excision of aganglionic bowel in Hirschsprung's disease. The limits of resection are defined by demonstrating ganglia in normal bowel; rapid frozen section of full thickness biopsies are often used for this purpose. Any difficulty in identifying ganglia in cryostat sections may be alleviated by special stains, such as a modified methyl green-pyronin (G. Anderson, personal communication).

Respiratory Tract

Many lung tumours are pre-operatively diagnosed by bronchoscopic biopsy or inter-costal needle biopsy. However, some tumours are relatively inaccessible, so tissue diagnosis must await thoracotomy. Squamous-cell carcinoma is fairly straightforward if well differentiated, as is oat-cell carcinoma in an uncrushed sample. With adeno-carcinomas one must be conscious of the possibility that the lesion may be a metastasis from some distant site. This may be academic if the lesion appears to be solitary because some non-pulmonary adenocarcinomas (e.g. from kidney) do well if these are excised; the primary can be dealt with later. Alternatively, the patient may be given the benefit of the doubt and pneumonectomy or lobectomy done in expectation that the lesion will eventually prove to be a primary.

The main pitfall in frozen sections of lung is the danger of interpreting organising pneumonia as a malignant neoplasm. Fibrosis and inflammation mimic stromal reactions to cancer; surviving swollen and distorted alveolar epithelium may be wrongly interpreted as neoplastic.

Urinary Tract

Nephrectomy for transitional-cell neoplasms of the renal pelvis involves removal of the kidney in continuity with a length of ureter. Because of the risk of multiple tumours or in situ malignancy it is often necessary to examine the ureteric resection margin.

The histological assessment of malfunctioning transplanted kidneys is often done by open biopsy when the organ is surgically explored. Obstruction, rejection, or vascular problems are the usual explanation for the unsatisfactory performance of the kidney and these can, with experience, be distinguished. If rejection is present it may be possible, again with experience, to say whether the rejection is predominantly of cellular or humoral type and give some indication as to the likelihood that the kidney may be salvaged by further immunosuppression.

Bone and Soft Tissues

Diagnosis of bone lesions by cryostat section is virtually impossible if the tissue is heavily mineralised. Softer tissue can be cut. No patient should undergo such a mutilating procedure as limb amputation for a tumour unless an absolutely unequivo-cal diagnosis of primary malignancy has been made histologically. (We must leave aside, for these purposes, whether amputation is the approriate treatment for any particular primary malignant bone tumour.) There is no evidence that incisional biopsy of osteogenic sarcoma adversely affects survival, unless definitive surgery is unduly delayed (Broström 1979). If a definite diagnosis of primary malignancy cannot be made on frozen sections, the surgeon must be told to await the interpretation of paraffin sections.

Frozen sections are usually requested on soft-tissue lesions either when it is neces-sary to determine the adequacy of resection or when a totally extirpative procedure must be justified by a previously unobtainable tissue diagnosis. The latter situation may arise when the lesion is in a relatively inaccessible location (e.g. retroperitoneum) or when a limb amputation is contemplated. The prospect of limb amputation reinforces the need for a confident diagnosis of primary malignancy.

Skin

Frozen sections are usually indicated only in suspected malignant melanoma, but this is practised in very few institutions. Various lesions can simulate the gross appearances of a malignant melanoma and some of these require just local removal with a fairly narrow margin. Melanoma has a notable propensity for local recurrence which can be curtailed by wide excision of the lesion after the diagnosis has been confirmed by cryostat section (Little and Davis 1974; McGovern 1976).

The suspected melanoma should be locally excised with just a narrow rim of skin and the frozen section then performed. If the lesion proves to be unequivocally a melanoma, a wider excision is done. If the lesion is minute, it is best to process the biopsy for paraffin sections rather than risk losing some or all of the lesion by cutting frozen sections. Never sacrifice a tiny biopsy on the chuck of a cryostat for the sake of expediency.

Reliability

Errors can occur in all histopathological diagnostic procedures owing to sampling, technique, communication and interpretation. Sampling is dealt with in Chap. 2. There is no doubt that considerable technical expertise is essential for the preparation of frozen sections of diagnostic quality; indifferent sections cause the diagnosis to be deferred until paraffin sections can be prepared. Communication errors are fortunately not common. Ideally the report should be written and legible. If telephoned, the message should be taken in theatre by a senior member of the nursing staff or a member of the surgical team.

Errors of interpretation are discovered by comparing the frozen section diagnosis with the final diagnosis, usually made on paraffin sections. Frozen sections should therefore always be filed for future reference. False-negative diagnoses are relatively inconsequential, except for the delay and inconvenience of the second operation. False-positive diagnoses of cancer are very serious because radical surgery may be done unnecessarily for benign lesions.

There was a tendency in the 1920s and 1930s for frozen sections to be unfavourably regarded as a diagnostic method; this was the inevitable consequence of erroneous diagnoses in some hands. Ewing (1925) became very reluctant to use rapid frozen sections.

Having made more errors by the frozen section method in breast cases than by gross examination, I have not resorted to frozen sections in this field for many years. . . . The cancer surgeon should become highly proficient in the recognition of cancer by sight and touch. No aid from frozen sections can replace this capacity.

Most surgical pathologists would, today, be horrified by this attitude. Even Joseph Bloodgood, formerly a staunch advocate of frozen sections, became more cautious in later years (Bloodgood 1934). Simpson (1937) claimed that enthusiasm for frozen sections was more dramatic than therapeutic.

What better impression can the surgeon make upon visiting practitioners . . . than to toss a specimen to a waiting pathologist and await his return a few minutes later, often out of breath, to give the diagnosis.

Improvements in section quality and wider recognition of difficult lesions ensure that the views promulgated by Ewing and some of his contemporaries did not gain long-lasting acceptance.

Accuracy rates in modern series range from 94.1% in 490 gynaecological cases (DiMusto 1970) and 93.4% in 212 breast specimens (Jennings and Landers 1957) to 98.34% in a large series of 2665 assorted frozen sections (Saltzstein and Nahum 1973). In this last series there were, however, four false-positive diagnoses of cancer.

Diagnostic reliability depends, at least in part, on active and continuous familiarity with the method, proficient technical staff, and surgeons with whom one has good rapport. Diagnostic errors can be minimised both by experience and by seeking a consensus with accessible colleagues. Pathologists, however senior, who do not regularly see frozen sections should probably not do them at all, even if they are thoroughly competent interpreters of paraffin sections.

For those who seek wide-ranging authoritative accounts of frozen section experience, I recommend the articles by Ackerman and Ramirez (1959) and Saltzstein and Nahum (1973). But Ernest Shaw will have the last word.

The mechanical processes involved in freezing, cutting, mounting and staining of fresh tissue sections are very simple and may be acquired by anyone after a little practice. The interpretation of the pictures presented under the powers of the microscope, however, is a different matter; here the pathologist is put on his mettle (Shaw 1923).

References

Ackerman LV, Ramirez GA (1959) The indications for and limitations of frozen section diagnosis: A review of 1269 consecutive frozen section diagnoses. Br J Surg 46: 336–350

Azzopardi JG (1979) Problems in breast pathology. Saunders, London Philadelphia Toronto, pp 2–7

Bloodgood JC (1934) Biopsy. Am J Surg 24: 331–344

Brostrom LA (1979) On the natural history of osteosarcoma: Aspects of diagnosis, prognosis and endocrinology. Acta Orthop Scand [Suppl] 183

Castleman B, Roth SI (1978) Tumors of the parathyroid glands. Armed Forces Institute of Pathology, Washington DC, p 64

Chang JP, Russell WO, Moore EB, Sinclair WK (1961). A new cryostat for frozen section technique. Am J Clin Pathol 35: 14–19

Cullen TS (1895a) A rapid method of making permanent specimens from frozen sections by the use of formalin. Bull Johns Hopkins Hosp 6: 67

Cullen TS (1895b) Beschleunigtes Verfahren zur Farbung frischer Gewebe mittelst Formalins. Zentralbl Allg Pathol 6: 448–450

Dahlin DC (1980) Seventy-five years' experience with frozen sections at the Mayo Clinic. Mayo Clin Proc 55: 721–723

DiMusto JC (1970) Reliability of frozen sections in gynaecologic surgery. Obstet Gynecol 35: 235–240

Duray PH, Flannery B, Brown S (1981) Tuberculosis infection from preparation of frozen sections. N Engl J Med 305: 167

Ewing J (1925) The diagnosis of cancer. JAMA 84: 1–4

Jennings ER, Landers JW (1957) The use of frozen sections in cancer diagnosis. Surg Gynecol Obstet 104: 60–62

Kanthack AA, Strangeways Pigg T (1897) Boiling water as a fixative and hardening agent: the revival of an old histological method for rapid diagnosis. Trans Pathol Soc London 48: 279–282

King DT, Hirose FM (1979) Chief cell intracytoplasmic fat used to evaluate parathyroid disease by frozen section. Arch Pathol Lab Med 103: 609–612

Little JH, Davis NC (1974) Frozen section diagnosis of suspected melanoma of the skin. Cancer 34: 1163–1172

Loquvam GS, Russell WO (1950) Accessory pancreatic ducts of the major duodenal papilla: normal structures to be differentiated from cancer. Am J Clin Pathol 20: 305–313

McGovern VJ (1976) Malignant melanoma: clinical and histological diagnosis. Wiley, New York London Sydney Toronto pp 131–145

MacMahon HE, DelVecho SB (1944) A simple technic of rapid sectioning. Engl J Med 231: 794

Mayo WJ (1929) The diagnostic value of microscopic examination of fresh frozen tissue. Surg Gynecol Obstet 49: 859–860

Nakazawa H, Rosen P, Lane N, Lattes R (1968) Frozen section experience in 3000 cases: Accuracy, limitations and value in residency training. Am J Clin Pathol 49: 41–51

Pick L (1987) A rapid method of preparing permanent sections for microscopical diagnosis. Br Med J I: 140–141

Saltzstein SL, Nahum AM (1973) Frozen section diagnosis: Accuracy and errors, uses and abuses. Laryngoscope 83: 1128–1143

Shaw EH (1910) The immediate microscopic diagnosis of tumours at the time of operation. Lancet II: 939–942

Shaw EH (1923) The immediate microscopic diagnosis of tumours at the time of operation. Lancet I: 218–223

Simpson WM (1937) The frozen section fetish. Am J Clin Pathol 7: 96–102

Sparkman RS (1962) Reliability of frozen sections in the diagnosis of breast lesions. Ann Surg 155: 924–934

Spjut HJ, Ramos AJ (1957) An evaluation of biopsy-frozen section of the ampullary region and pancreas: a report of 68 consecutive patients. Ann Surg 146: 923–930

Terry BT (1927) A new and rapid method of examining tissue microscopically for malignancy. J Lab Clin Med 13: 550–560

Wang CA, Rieder SV (1978) A density test for intraoperative differentiation of parathyroid hyperplasia from neoplasia. Ann Surg 187: 63–67

Wright JR (1985) The development of the frozen section technique, the evolution of surgical biopsy, and the origins of surgical pathology. Bull Hist Med 59: 295–263

9 Diagnostic Electron Microscopy

Although the impact of electron microscopy on biomedical research is undeniable, diagnostic applications have only recently become well established. The science of electron optics led to the first practical electron microscopes in the 1930s and published papers describing the ultrastructure of biological specimens were almost commonplace by the 1950s (Peven and Gruhn 1985). Originally it was assumed that fastidious attention to fixation was imperative and that specimens which had been treated suboptimally were useless. It gradually became clear, however, that diagnostically useful information could be obtained from tissues that had first been routinely handled for light microscopy. The technique is now regarded as an indispensable adjunct to light microscopy for the interpretation of renal biopsies and the histogenetic diagnosis of tumours.

Principles of Transmission Electron Microscopy

The Electron Microscope

The resolution of an optical system is limited by the nature of the beam that is used to form an image of the object. The relatively high resolution of the electron microscope, about 0.2 nm as compared with 200 nm for the light microscope, is due to the physical properties of the electron itself.

The basic pattern of an electron microscope is very similar to that of a light microscope, with condenser, specimen, objective lenses, etc. (Fig. 9.1). However, electrons will not pass through air or glass to any appreciable extent. The electron source is an electrically heated filament placed at one end of an evacuated cylindrical metal column. The whole electron gun is held at a high negative charge (voltages from -20 to -100 kV are typical) relative to the rest of the column which is earthed. The electron beam is finely collimated and accelerated down the column by the potential difference between the electron gun and the anode, the latter connected to the

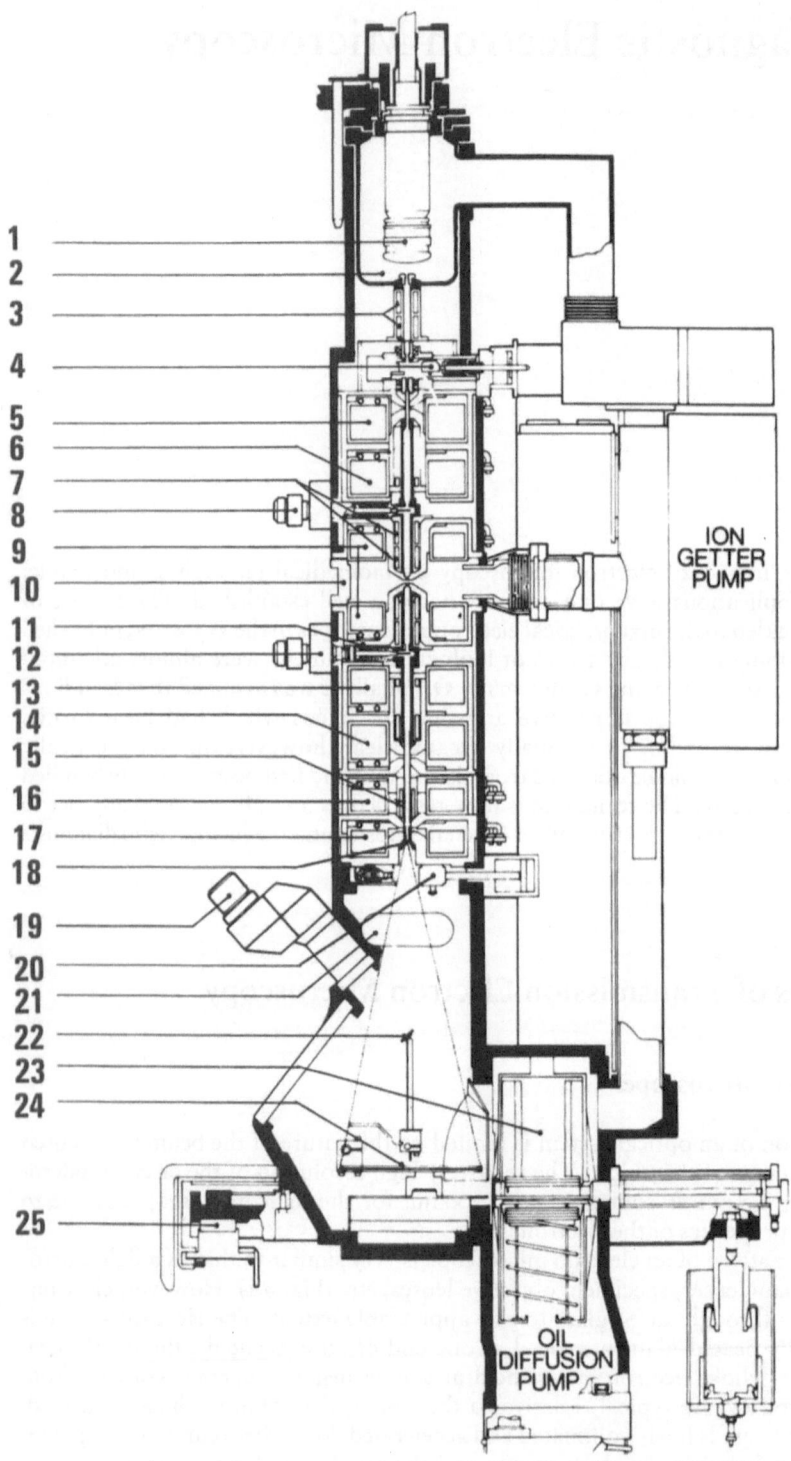

earthed column. Near the middle of the column is located the specimen, a thin section supported on a metal grid attached to a stage which can be moved through fine micrometer controls. Electromagnetic lenses magnify and focus the image formed by the electrons that traverse the specimen. A visible image is either formed on a phosphor screen or recorded on a photographic emulsion. Different instruments differ in detail. Some are highly sophisticated with a resolving power close to the physical limits. Magnification may range up to × 500 000. For diagnostic purposes most clinically significant information can be picked up below × 20 000 magnification.

Specimen Processing

Tissue for electron microscopy should ideally be obtained as fresh as possible and gently cut with a clean sharp blade into small (approximately 1 mm) cubes or thin slices and fixed in glutaraldehyde, osmium tetroxide, or some other fixative specially devised for ultrastructural work. Because of sampling problems, extra care must be taken to ensure that the lesion or tissue is adequately represented in the material taken for electron microscopy. Adjacent slices can be examined by routine light microscopy to help select the right areas.

In many instances the need to do electron microscopy arises only after light microscopy has been done; the tissue has then spent several days in formalin. Ashworth and Stembridge (1964) showed, however, that formalin-fixed biopsy and autopsy specimens can be processed for electron microscopy and still give useful results. Tissue can even be removed from paraffin wax blocks (Johannessen 1977) or from selected areas of a paraffin wax section (Rossi et al. 1970). The results can be disappointing, but the resistance of organelles, membranes, desmosomes, fibres and organisms to such rough treatment is quite remarkable.

After fixation the tissue block can be post-fixed in osmium tetroxide to enhance the electron density of membranes and other components. The tissue is embedded in epoxy resins, after dehydration by passage through alcohol, and sections cut with glass or diamond knives. "Semi-thin" (approximately 1 μm) sections are initially cut, mounted on glass slides and stained, usually with toluidine blue. These sections are examined by light microscopy to ensure that the tissue block is representative and free from undesirable necrosis or distortion. The exact area required can be located and the block trimmed so that this area is included on the final section. Thin sections (approximately 80 nm) are then cut, mounted on metal grids, usually of copper, stained to increase contrast, and inserted through an air-lock into the specimen chamber of the electron microscope. Most stains are salts of lead or uranium; these have an affinity for membranes and proteins rendering them electron opaque. Other stains can be used to show specific components (e.g. ruthenium red for the glycocalyx). Immunolocalisation can be done with antibodies conjugated either to ferritin or gold,

Fig. 9.1. Diagrammatic column section of the Philips EM 400 transmission electron microscope. Key to principal components: 1, electron gun; 2, anode; 3, gun alignment coils; 4, gun air-lock; 5, 1st condenser lens; 6, 2nd condenser lens; 7, beam tilt coils; 8, condenser aperture; 9, objective lens; 10, specimen block; 11, stainless-steel bellows; 12, diffraction aperture; 13, diffraction lens; 14, intermediate lens; 15, column-lining tubes; 16, 1st projector lens; 17, 2nd projector lens; 18, 200 μ vacuum differential diaphragm; 19, × 12 binocular; 20, column vacuum block; 21, 35 mm roll-film camera; 22, focusing screen; 23, plate camera; 24, main screen; 25, anti-vibration mounts. (Courtesy of Philips, Eindhoven)

which are inherently electron dense, or to peroxidase, which reacts with diaminobenzidine to give an oxidised osmiophilic product (see p. 68).

Examination of the section is done by viewing the image on the phosphor screen. Electron micrographs are, however, essential. Not only do they give a permanent record of the ultrastructure, but more detail can be seen than on the screen. The photographic documentation should always include representative low-power (approximately ×1500) views even when no obvious noteworthy features can be picked up on screening at this magnification. These low-power micrographs often help to put those taken at a higher magnification in proper context.

The entire exercise, from receipt of the tissue to the production of sections for examination in the microscope, can be done at least as rapidly as paraffin-section processing (Rowden and Lewis 1974).

Diagnostic Value of Electron Microscopy

Despite the growth of interest in immunohistology as an adjunct to routine light microscopy, electron microscopy finds increasingly wide acceptance as a diagnostic tool in histopathology (Williams and Uzman 1984). A comprehensive survey of diagnostic applications is not possible here, so it will suffice to explain briefly how the technique has helped to investigate selected diagnostic problems.

Renal Disease

The undisputed role of electron microscopy in the diagnosis of glomerular disease has been reviewed by Spargo (1975). He claims, as do others, that routine electron microscopy on renal biopsies minimises the need for special stains. Subtle changes, only inferred from the light microscopic appearances, can be directly visualised with the electron microscope. Epithelial cell foot-process fusion, membrane thickening, immune-complex and amyloid deposition, mesangial proliferation can all be clearly discerned (Fig. 9.2). Only minute deposits of complexes are necessary to cause abnormal permeability of the glomerular basement membrane resulting in clinically significant proteinuria; these can escape detection by the light microscope, as in the "minimal change" lesion.

Tumour Histogenesis

Rosai and Rodriguez (1968) were among the first to use electron microscopy to solve practical diagnostic problems in oncology. The principle is now well established and its usefulness is subject to continuing exploration (Gyorkey et al. 1975; Carr and Toner 1977; Ghadially 1980; Henderson and Papadimitriou 1982). Even tissue fixed in acid non-buffered formalin and salvaged from a wax block may contain ultrastructurally recognisable histogenetic markers.

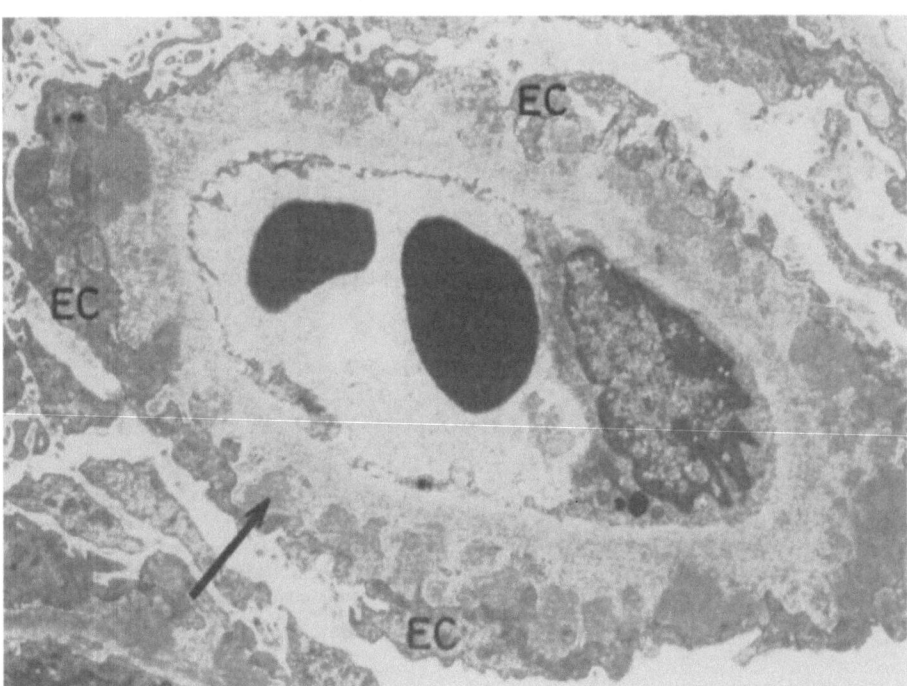

Fig. 9.2. Deposits of immune complex (*arrowed*) on the glomerular basement membrane in glomerulonephritis. Epithelial cell (*EC*) foot-process fusion is also evident. Transmission electron micrograph, × 2000

The electron microscope is not a good tool for distinguishing benign from malignant neoplasms nor, for that matter, helpful in distinguishing between neoplastic and non-neoplastic lesions. No consistent or reliable ultrastructural hallmarks of neoplasia or malignancy have yet been described. The purpose of doing electron microscopy on a neoplastic lesion is to identify ultrastructural features of cellular differentiation, such as desmosomes and tonofilaments in squamous-cell carcinoma or melanosomes in malignant melanoma (Figs. 9.3–9.6). These and other histogenetic markers are listed in Table 9.1.

When examinining neoplastic tissue one must be certain that the actual neoplastic cell population is being seen, rather than stromal elements or remnants of infiltrated normal tissue. Also, just because a cell contains something does not mean that that cell has actually made it. Melanosomes are donated to the basal cells of the normal epidermis and may therefore be also found in basal-cell papillomas and carcinomas. Fat may be ultrastructurally found in many neoplastic cells and is not a reliable marker of adipose differentiation.

The frequency with which a particular feature occurs in a tumour is important. A solitary desmosome or desmosome-like structure is not sufficient evidence of epithelial differentiation for example. A considerable amount of aberrant biological behaviour occurs in neoplasia; it is important to distinguish really significant features from those that are mere epiphenomena.

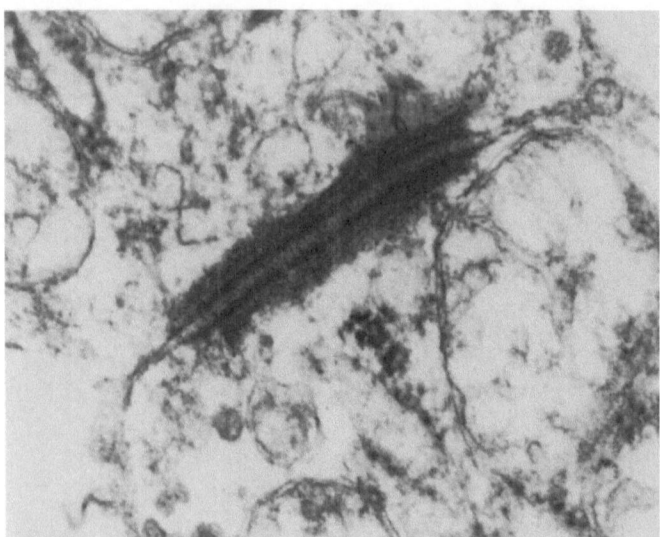

Fig. 9.3. Well-preserved desmosome between adjacent cell membranes in a skin tumour. Tissue had been routinely fixed in formalin. Abundant tonofilaments also found; these appearances favour a squamous histogenesis. Transmission electron micrograph, × 50 000

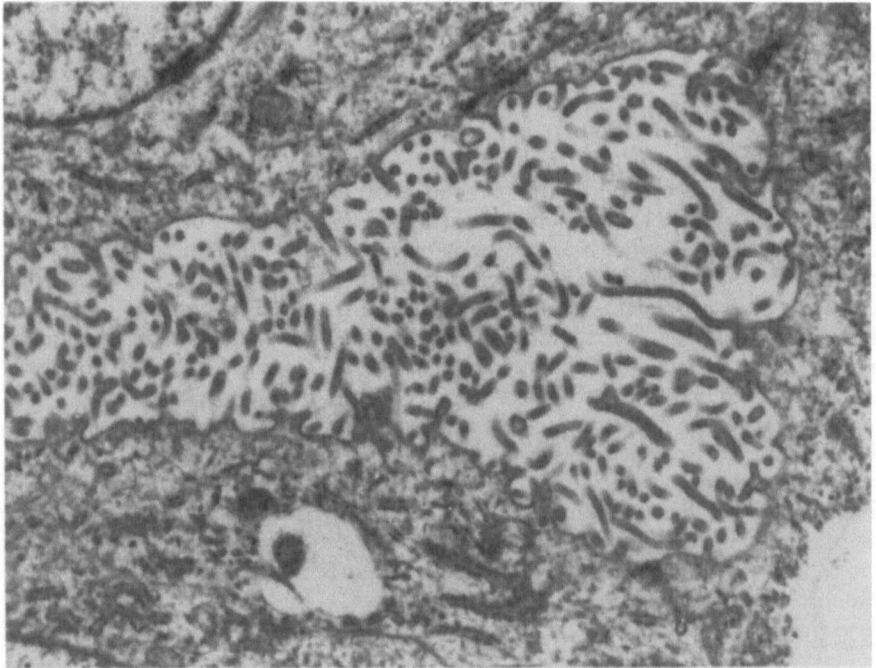

Fig. 9.4. Irregular long busy microvilli help to distinguish this peritoneal mesothelioma from an adeno-carcinoma. The tissue had initially been routinely fixed in formalin. Transmission electron micrograph, × 10 000

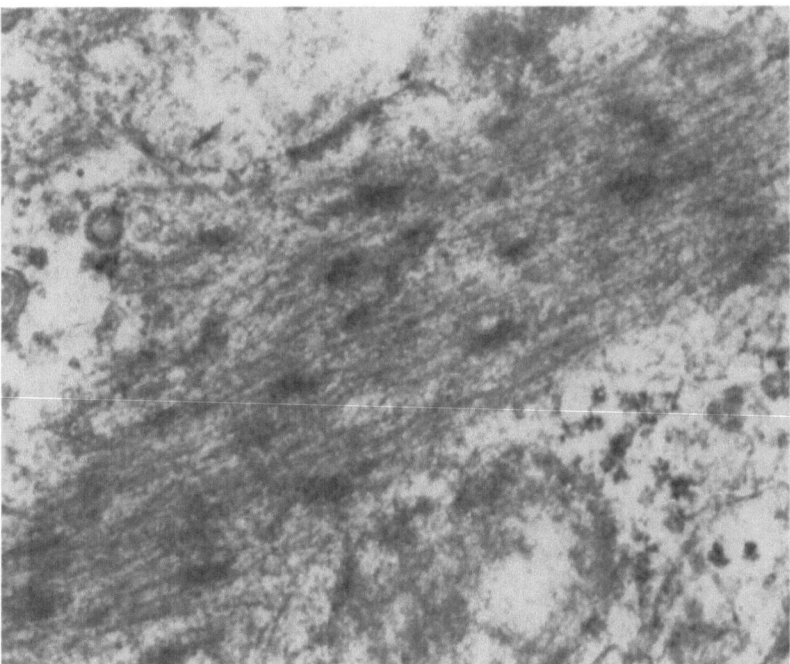

Fig. 9.5. A thick bundle of intracytoplasmic filaments with focal dense bodies, a prominent feature in this formalin-fixed vulval tumour. The appearances indicate smooth-muscle differentiation. Transmission electron micrograph, × 40 000

Virus Identification

Viruses can be identified by electron microscopy in either thin sections or negatively stained preparations. In histopathological diagnostic work it is usually only the former variety of material that is available. Viruses have distinctive morphological characteristics that enable them to be broadly identified; size, shape and internal structure are the major distinguishing features (Yunis et. al. 1977). The location within a tissue or cell is also important since many viruses exhibit a highly specific propensity to infect certain cells and not others, and some reside exclusively in either the cytoplasm or nucleus (Fig. 9.7).

Miscellaneous Applications

Only a limited survey is possible and some of the applications cited below might justifiably be thought of as worthy of consideration in greater depth.

Amyloid stains such as Sirius red and thioflavine T are often found to give inconclusive results; electron microscopy is often done to resolve the problem. Fortunately, the ultrastructure of amyloid seems to withstand formalin fixation and even wax embedding reasonably well. It has a characteristic microfibrillar appearance irrespective of its aetiology (e.g. primary, secondary, APUD amyloid).

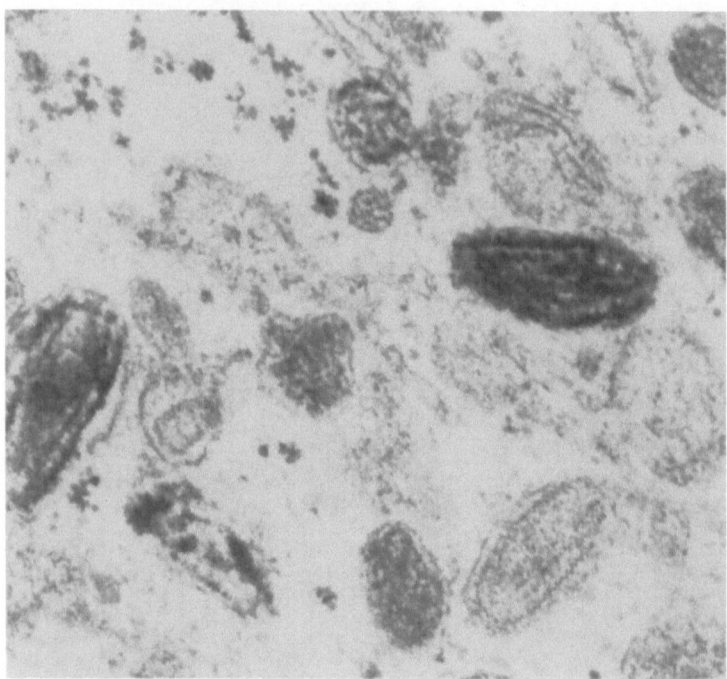

Fig. 9.6. Melanosomes in a malignant melanoma on the leg. The internal structure of the melanosome alters as the organelle matures and accumulates pigment. Formalin fixation. Transmission electron micrograph, × 60 000

Table 9.1. Some ultrastructural markers of tumour histogenesis[a]

Morphological feature	Possible significance
Cell surface	
Junctional complexes (other than true desmosomes)	Epithelium, muscle, nerve sheath, mesothelium, etc.
Tight junctions	Possibly glandular epithelium
Desmosomes	Epithelium, squamous if abundant
Short regular microvilli	Glandular epithelium
Long irregular microvilli	Mesothelial lesion
Irregular ruffled membrane	Possibly histiocytic
Cytoplasmic	
Tonofilaments	Epithelial, probably squamous if very abundant
Cross-striated thick and thin myofilaments	Skeletal muscle lesion
Longitudinal myofilaments with focal dense bodies	Smooth muscle, myoepithelium, myofibroblasts
Copious rough endoplasmic reticulum	Plasma-cell lesion
Large glycogen aggregates	Ewing's tumour
Mitochondria with tubular cristae	Steroid-producing cell
Intracytoplasmic acini	Aenocarcinoma, probably lobular carcinoma of breast
Ribosome-lamella inclusions	Hairy-cell leukaemia

Table 9.1 *(continued)*

Morphological feature	Possible significance
Langerhans' granules	Histiocytosis
Secretory granules	
Mucin (pleomorphic)	Mucin-secreting epithelium
Zymogen (uniform electron-dense, 250–500 nm diameter)	Exocrine (e.g. pancreas, salivary)
APUD (uniform electron-dense, 100–400 nm diameter)	APUD cell lesion (e.g. carcinoid, islet-cell tumour, oat-cell carcinoma)
Melanosomes	Probable melanocytic lesion
Weibel-Palade bodies	Vascular endothelium
Nucleus	
Deeply invaginated or crenated profile	Smooth muscle cell or Sézary cell
Stroma	
Amyloid fibrils	Probable APUDoma
Luse bodies (long-spacing collagen)	Probable schwannoma

ᵃ Diagnosis is rarely possible from the presence of one feature alone. Ultrastructural findings should always be interpreted in the context of the gross and light microscope appearances.

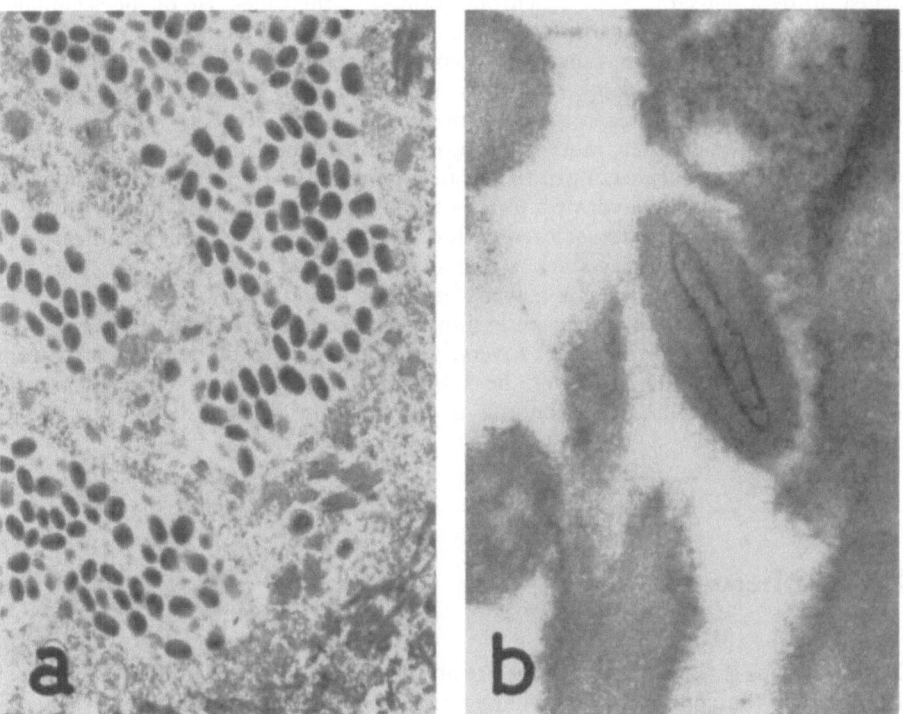

Fig. 9.7a,b. Virus particles. a Molluscum contagiosum showing numerous intracytoplasmic viruses of the pox virus group. Formalin fixation. Transmission electron micrograph, × 10 000. b Ecthyma contagiosum (orf) on the finger of an abattoir worker. Virus particles were easily found in the formalin-fixed tissue retrieved from a paraffin wax block. Transmission electron micrograph, × 125 000

Certain metabolic disorders have distinctive ultrastructural features, such as those associated with storage diseases (Ishak et al. 1978).

The ultrastructural identification of organisms other than viruses is a useful adjunct to routine light microscopy, such as in Whipple's disease.

Recently, ultrastructural lesions in cilia, such as absence of dynein arms in Kartagener's syndrome, have been reported and it is likely that increasing attention will be paid to subtle abnormalities such as these (Eliasson et al. 1977).

Scanning Electron Microscopy

The principle of scanning electron microscopy differs from that of transmission electron microscopy. Instead of the electron beam passing through the specimen, usually a thin section, and forming an image of its interior, an image of the specimen's surface is obtained. The instrument consists of an evacuated chamber in which a fine beam of electrons is scanned across the object by electromagnetic lenses; the scan pattern is similar to the raster of a television screen. Electrons reflected from the surface of the specimen, usually tilted at an angle of about 45° to the beam, are picked up by a detector. After synchronisation with the time-base of the scanning beam, the resulting image is displayed on a cathode-ray tube where it can be photographed. The magnification ranges up to approximately × 50 000 with a resolution of less than 10 nm. The depth of field is remarkable and enables extremely crisp photomicrographs of microanatomical structures to be obtained.

Free cells, tissue cultures, mucosal surfaces, skin surfaces and structures exposed on the surface of cleanly sectioned tissues (e.g. glomeruli, alveoli) can be examined. The specimen must be dehydrated, usually through graded alcohols, and dried; the most satisfactory results are achieved with critical-point drying from liquid carbon dioxide. The specimen is then stuck to a small metal stub with a conducting adhesive and coated in a vacuum with a thin layer of gold and palladium.

Diagnostic applications are currently limited. The undeniable beauty of scanning electron micrographs is sometimes inversely proportional to the amount of useful information they contain. However, the technique is now producing clinically useful information (Fig. 9.8) and will continue to be subject to further study (Carr et al. 1980).

X-Ray Microanalysis

When electrons strike an object, some are absorbed and energy is emitted as X-rays. The energy of the emitted X-rays is characteristic for each element, so that the elemental composition of the object can be deduced if the energy peaks are displayed on a calibrated scale. This technique, X-ray microanalysis, is therefore an adjunct to electron microscopy; the apparatus can be coupled to suitable transmission or scanning electron microscopes.

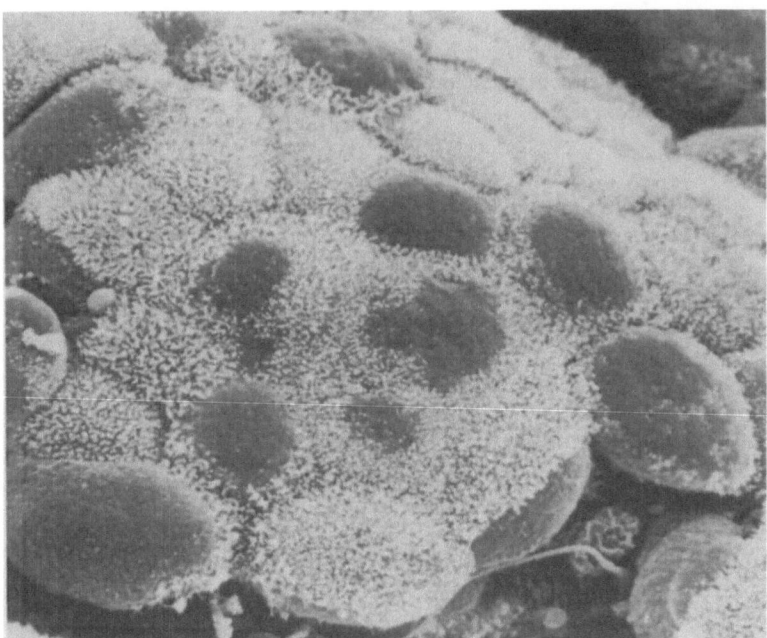

Fig. 9.8. Scanning electron micrograph of urothelium from a human bladder bearing a transitional-cell carcinoma. The cells are partially or completely covered with microvilli. Bladders showing this microvillous change are more prone to develop recurrent tumours. (Newman and Hicks 1977) Scanning electron micrograph, × 2400

Tissues and sections should ideally be unstained to avoid introducing contaminants into the specimen, but of course it does mean that the transmission image lacks contrast. Standard copper grids can be used except when seeking copper itself, or elements that give peaks close to it, in the specimen; nickel grids can be used as an alternative.

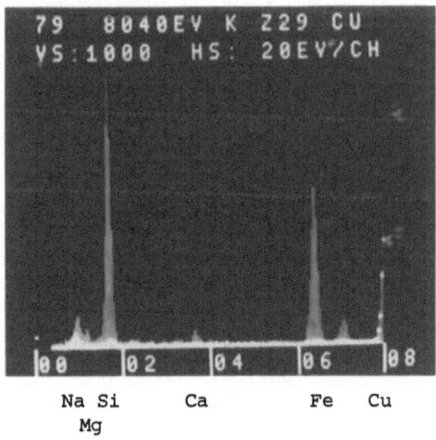

Fig. 9.9. Identification of an asbestos particle in lung by X-ray microanalysis. Energy peaks are due principally to silicon and iron with sodium, magnesium and calcium in lesser quantities. Copper peak on extreme right is from grid supporting the section

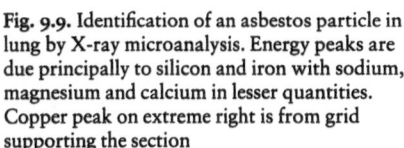

Diagnostic applications include the detection and ultrastructural localisation of copper in Wilson's disease and some other hepatic disorders, silver in cases of suspected argyria, and silicon and associated elements in asbestosis and other pneumoconioses (Fig. 9.9).

References

Ashworth CT, Stembridge VA (1964) Utility of formalin-fixed surgical and autopsy specimens for electron microscopy. Am J Clin Pathol 42: 466–480

Carr I, Toner PG (1977) Rapid electron microscopy in oncology. J Clin Pathol 30: 13–15

Carr KE, McLay ALC, Toner PG, Chung P, Wong A (1980) SEM in service pathology: a review of its potential role. In: Johari O, Becker RP (eds) Scanning electron microscopy/1980/III. Scanning electron microscopy Inc., O'Hare, Ill, pp 121–138

Eliasson R, Mossberg B, Cammer P, Afzehius BA (1977) The immotile-cilia syndrome: a congenital ciliary abnormality as an etiologic factor in chronic airways infections and male sterility. N Engl J Med 297: 1–6

Ghadially FN (1980) Diagnostic electron microscopy of tumours. Butterworths, London Boston

Gyorkey F, Min KW, Krisko I, Gyorkey P (1975) The usefulness of electron microscopy in the diagnosis of human tumours. Hum Pathol 6: 421–441

Henderson DW, Papadimitriou JM (1982) Ultrastructural appearances of tumours: a diagnostic atlas. Churchill Livingstone, Edinburgh London Melbourne New York

Ishak KG, Edwards RH, Schochet SS (1978) Pathology of inborn errors of metabolism. In: Anthony PP, Woolf N (eds) Recent advances in histopathology, no. 10. Churchill Livingstone, Edinburgh London New Yor, pp 91–137

Johannessen JV (1977) Use of paraffin material for electron microscopy. Pathol Annu 122: 189–224

Newman J, Hicks RM (1977) Detection of neoplastic and preneoplastic urothelia by combined scanning and transmission electron microscopy of urinary surface of human and rat bladders. Histopathology. 1: 125–135

Peven DR, Gruhn JD (1985) The development of electron microscopy. Arch Pathol Lab Med 109: 683–691

Rosai J, Rodriguez HA (1968) Application of electron microscopy to the differential diagnosis of tumours. Am J Clin Pathol 50: 555–562

Rossi GL, Luginbuhl H, Probst D (1970) A method for ultrastructural study of lesions found in conventional histological sections. Virchows Arch (Pathol Anat) 350: 216–224

Rowden G, Lewis MG (1974) Experience with a three-hour electron microscopy biopsy service. J Clin Pathol 27: 505–510

Spargo BH (1975) Practical use of electron microscopy for the diagnosis of glomerular disease. Hum Pathol 6: 405–420

Williams MJ, Uzman BG (1984) Uses and contributions of diagnostic electron microscopy in surgical pathology. Hum Pathol 15: 738–745

Yunis EJ, Hashida Y, Haas JE (1977) The role of electron microscopy in the identification of viruses in human disease. Pathol Annu 121: 311–330

10 Quantitative Methods

Quantitative methods have an assured role in histopathological research, where the requirement for the numerical expression of morphological information is necessary to reduce observer bias and to permit the application of statistical methods for the analysis of results. There is, however, an increasing interest in the application of quantitative methods to the solution of diagnostic problems in clinical biopsies (Baak et al. 1982). Quantitative methods may be advantageous in the following situations:

—To make the diagnosis in cases in which the structural features of the presumed diseases deviate only slightly from the normal state and may therefore be easily overlooked
—To assist in the distinction between two or more diseases having morphological resemblances
—To improve the objective documentation of lesions
—To compare repeat biopsies accurately in order to assess therapeutic responses
—To enable the performance of statistical analyses

Many medical practitioners, irrespective of their specialty, are still somewhat resistant to the idea of introducing quantitative methods and computer analysis into clinical work, which they regard as an inductive activity—almost an art-form (Baak et al. 1982). But the effect can be liberating, enabling the practitioner to further develop the subject, particularly if the quantitative analyses can be automated thus releasing personnel to concentrate on other work of a more inductive or subjective nature.

Stereological Principles

Some quantitative procedures are simple and need no further elaboration. These include determining the weight and linear dimensions of resected specimens and lesions. Stereology, the derivation of quantitative information about the three-dimensional composition of an object, is a more complex procedure. (The word "morphometry" is often used interchangeably with "stereology" though the two terms are not strictly synonymous. Morphometric methods are for the derivation of purely *two-dimensional* information (e.g. length, area) about a tissue from cut surfaces or sections).

Stereological techniques were originally devised by geologists for rock analysis. Delesse, a French geologist, enunciated a fundamental stereological principle in 1847. The Delesse principle states that the volumetric composition of a material composed of two or more randomly distributed and structurally separate components can be estimated from area measurements on random plane surfaces or sections of negligible thickness; the area of a surface or section occupied by each randomly distributed component is proportional to the volume fraction of that component in the whole object. Modern methods for the stereological analysis of tissues are based on this principle.

Before dealing with stereological methods in detail it must be emphasised that all stereological and morphometric measurements yield estimates rather than absolute values. Statistical considerations must be applied because individual sections or cut surfaces are never absolutely representative of the tissue as a whole. Specimens should be sampled in such a way as to avoid bias introduced by consciously selecting favourable gross areas or microscope fields. Tissue blocks can be taken either by systematic sampling according to a predetermined protocol or by random sampling using a number grid superimposed on the object and a random number table. Single sections may be examined from multiple blocks prepared from large specimens; if the biopsy is small, sections taken at intervals through the block will be needed. Assuming that histological methods are used, the next problem is to find the minimum number of random microscope fields that must be assessed to achieve a result of acceptable accuracy; the limits of acceptable accuracy for biological work are generally held to be $\pm 5\%$ of the actual value. To find the number of random fields necessary to give a result of acceptable accuracy, a cumulative mean or summation average graph is constructed in which the arithmetic mean of the quantitated feature is plotted against the number of fields counted; the mean is calculated after each field has been assessed. The plot deviates wildly for the first few fields, but gradually settles down until it becomes almost a straight line. At this point, taken to be the mean, $\pm 5\%$ of the mean is calculated and the plot is examined to determine when the cumulative mean ceased to fluctuate beyond those limits. The number of fields needed to achieve this level of accuracy is then read off the graph. The adequacy of gross sampling can be similarly determined.

Stereological Methods

Stereological methods can be applied to gross specimens and light microscope and electron microscope images. The principles and rules remain the same although correction factors to compensate for tissue shrinkage may have to be used where appropriate. The geometrical form of the object may be such as to necessitate variation from standard stereological procedures (Underwood 1972; Mayhew and Cruz Orive 1974).

Volumetric Analysis

The volumetric composition of a tissue can be estimated according to the Delesse principle from area measurements of randomly distributed components in plane surfaces or infinitely thin sections. The volumetric composition is usually expressed as a porportion, ratio or percentage. If absolute values are required it is first necessary to determine the volume of the entire organ or lesion. This can be done by simple geometry if the object conveniently conforms to some standard geometrical shape such as a cube or sphere; this is not usually the case. Displacement methods can be used, but these may be difficult with small organs or with tissues like lung that deform under pressure. Alternatively the volume can be determined by applying Simpson's rule (Dunnill 1968). The lesion or organ is cut into n slices of equal thickness, h, and the area of one face of each slice, A_0, A_1, $A_2 \ldots$, is measured, by planimetry for example (see below). The volume V can be calculated as follows:

$$V = 1/3h[(A_0 + A_n) + 4(A_1 + A_3 + \ldots + A_{n-1}) + 2(A_2 + A_4 + \ldots + A_{n-2})]$$

Care must be taken to avoid shrinkage during fixation and processing; any that does occur must be measured so that the final results can be corrected.

Four suitable methods exist for measuring areas in surfaces or sections of tissues.

1. *Measurement of Traced Images.* The original method used by Delesse and his contemporaries to measure areas for volumetric rock analysis was to trace the outlines of the different components onto paper. The traced areas were then cut out and weighed. If the paper used is of uniform thickness the weight of each piece will be proportional to its area and hence, according to the Delesse principle, the volume of that traced component. This method has been successfully applied to biological problems, but it is laborious. An alternative to weighing is to trace the image onto squared paper; the number of squares coincident with each component will be a measure of the area. If paper-weighing or square-counting is used for histological work it is first necessary to trace the outlines of a projected image on the paper.

2. *Planimeter.* This mechanical device computes the area from angular movements of the arms of the instrument. A surface or projected image is prepared and the margins of each component profile are delineated with the movable arm of the planimeter. The area circumscribed by each movement is read off directly from the instrument.

3. *Linear Integration.* Rosiwal, also a geologist, discovered in 1898 that areas could be estimated from the fractional length of a series of lines that coincided with different components on the plane surface of a rock. In practice the lines were arranged as a grid for superimposition on the plane surface; the sum total length of the portions of lines coincident with a given component would be proportional to its area and hence its volume. For biological work the image can be projected onto a line-grid drawn on a screen or alternatively superimposed on the image by using a grid drawn on an eyepiece graticule. This method is cumbersome because of the necessity of measuring the length of that part of each line coinciding with a given component. However, the method can be adapted for biological work and the principle is used in some automatic scanning devices.

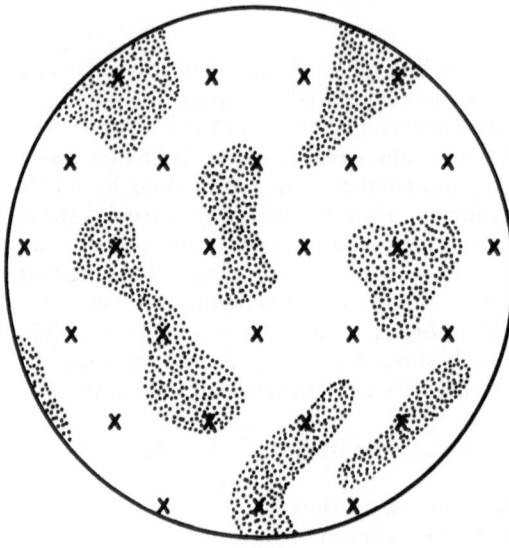

Fig. 10.1. Estimation of area by point-counting. The diagram represents a microscope field of a thin section containing a randomly distributed shaded component. A grid of points (*intersection of crosses*) is superimposed on the image (e.g. using an eyepiece graticule); the proportion of the total number of points coinciding with the shaded component (10 out of 27 points in this instance) gives an estimate of the relative area of the field occupied by the shaded component. This in turn, according to the Delesse principle, is proportional to the relative volume of the randomly distributed component in the tissue as a whole.

4. *Point-Counting*. This is the simplest of all methods for volumetric analysis by area measurements. It was devised in 1933 by Glagoleff, again for geological work. Glagoleff showed that areas can be estimated by superimposing a lattice of points over the plane surface or section. The proportion of points coincident with different component profiles is proportional to their area in the section and, from the Delesse principle, their volume in the object as a whole (Fig. 10.1). Chalkley (1943) was the first to use point-counting for biological work. This is now the most widely used of the non-automatic methods for area measurements in histopathological research and diagnosis.

For practical applications the lattice of points is either drawn on a screen to receive a projected image, on a transparent sheet for superimposition on the cut surface of a gross specimen, or incorporated as an eyepiece graticule in a microscope. The format of the point-lattice is not absolutely critical. It may be regular or random. There may be few points or many (Chalkley's had only five), but the cumulative means procedure must be used to determine the number of fields to be counted.

Surface Area

The surface area of randomly distributed structures in a tissue (e.g. alveolar wall, basement membranes) can be estimated by the method of mean linear intercept. A grid of lines of known length are either randomly superimposed on the object using an eyepiece graticule or drawn on a screen for projected histological images. The number of times, m, membrane profiles intercept the lines is recorded. If L is the total line length in the grid and N is the number of fields assessed, the mean linear intercept L_m is calculated as follows:

$$L_m = \frac{NL}{m}$$

If V is the volume of the organ or lesion in question, corrected for shrinkage if necessary, then the total surface area of the membranes is calculated as follows (Dunnill 1968):

$$\frac{2V}{L_m}$$

Surface-to-Volume Ratios

Surface-to-volume ratios can be derived by a combination of volumetry by point-counting and surface density estimation by mean linear intercept (Chalkley et al. 1949). The grid devised by Weibel is especially suited to this task (Weibel et al. 1966). As previously stated, the grid can be either used as an eyepiece graticule or drawn on a screen. This has been suggested as a suitable way to monitor changes in jejunal biopsies accurately (see pp. 171–172).

Particle Size and Density

Spherical particles are the easiest to measure in sections. Sections should ideally be at least as thick as the diameter of the structures to be measured, and the equator of the particle should be included in the plane of section. By focusing up and down through the plane of section the level of maximum diameter can be identified; the diameter at this level is measured with an eyepiece micrometer. The true diameter of spheres transected by thin sections, or exposed on plane cut surfaces, can be calculated from the mean diameter of the transections, d, as follows:

$$\frac{4d}{\pi}$$

Particle concentration in a given volume is difficult to estimate because a section of finite thickness contains either whole particles as well as portions of particles or only portions of particles if the section thickness is less than the particle diameter. Particle-counting in sections will therefore result in an overestimate of the true particle concentration. Weibel has proposed a solution to this problem which utilises a coefficient based on the shape of the particles; the particles must be randomly distributed and conform to a well-defined shape (Weibel 1963).

For authoritative accounts of mathematical desiderata, sampling for stereological analyses, and construction of grids, the reader is referred to the articles by Weibel and colleagues (Weibel 1963; Weibel et al. 1966; Weibel and Elias 1967) and by Dunnill (1968).

Automatic and Semi-automatic Image Analysis

Stereology by direct visual methods such as point-counting can be time-consuming, lead to observer fatigue and hence error. Automatic and semi-automatic image analysis machines use electronic methods to derive stereological information from tissues and cells; many of these machines were originally developed for metallurgical purposes.

For automatic image analysis the section, smear, etc. must be stained or treated in such a way that the machine can discriminate between the tissue components of interest by their monochromatic shade or "greyness" (grey value). Colour filters may be needed to enhance contrast. The image is picked up by a television camera and electronically analysed. These systems (e.g. Quantimet, Cambridge Analysing Instruments Ltd.) are extremely versatile and will compute a wide variety of stereological measurements. The main drawback for biological work is the problem of staining or treating the specimen so that the grey value of the component of interest is sufficiently different from the background of unwanted components. Semi-automatic systems in which the components of interest are delineated by the operator do not suffer from this drawback. Indeed, because the image produced by automatic analysers often requires so much manual editing, many users now prefer these rather less expensive semi-automatic systems.

For quantification on semi-automatic analysers (e.g. MOP, Kontron Messgeräte GmbH) there is no need for special staining except that which is necessary to enable the operator to identify the tissue components. The apparatus, like the automatic machines, is extremely versatile; it can be used with projected images, electron micrographs, gross tissues, and with an adapted light microscope by using a light-pen image superimposed in the focal plane of the eyepiece (Fig. 10.2).

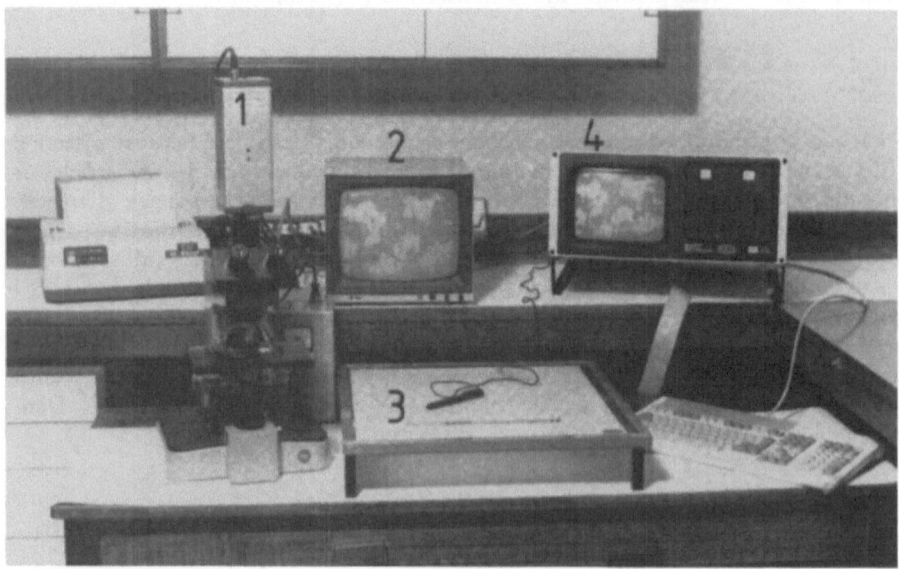

Fig. 10.2. Typical semi-automatic image-analysis system (MOP-Videoplan, Kontron). The section to be quantified is placed on the stage of a microscope fitted with a television camera (1). The operator delineates the relevant structures in the image on the monitor (2) by moving a cursor over the surface of the digitising tablet (3). The coordinates of the delineated image are analysed by the computer (4).

Microspectrophotometry

Quantification of the chemical composition of individual cells or their nuclei can be done after appropriate specific staining under well-controlled conditions followed by measurement of the optical density, or emitted light intensity in the case of fluorimetry, at defined wavelengths.

The most common application of these methods is the estimation of nuclear DNA content. Sections, imprints or smears are stained by the Feulgen method employing Schiff's reagent after controlled hydrolysis of the DNA in hydrochloric acid; the optical density of individual nuclei is measured at 550 μm wavelength using a microscope equipped with a sensitive photoelectric cell. Blood lymphocytes are often used as the standard reference control for the diploid DNA value. The DNA content can be expressed in arbitrary units calculated from the optical density (Böhm and Sandritter 1975). This has been largely superceded by flow cytometry.

Flow Cytometry

This is a relatively new technique for the rapid analysis of cellular parameters. The original concept is credited to Kamentsky et al. (1965) who devised a "rapid cell spectrophotometer", the first practical application of which was screening for cervical cancer (Koenig et al. 1966). Modern instruments, however, employ two important refinements attributed to Van Dilla et al (1969): the suspension of cells to be analysed is centred in the analysing light beam by being carried in a stream enclosed within a laminar-flowing liquid sheath; and coherent light from an argon-ion laser is used to generate the interrogating beam. The significance of these developments will be clarified by the following description of a modern flow cytometer equipped with a facility for cell sorting.

A suspension of isolated cells (solid tissues have to be enzymatically disaggregated before they can be analysed) is passed as a very fine stream through an interrogating laser light beam and the emitted fluorescence and the scattered or transmitted laser light from each cell is collected and its intensity measured and stored for analysis and display by a computer (Fig. 10.3). Cell size can be deduced from the forward angle light scatter; this is directly proportional to the size of the cell only if it were a perfect homogeneous sphere, which animal cells of course only approximate. The granularity of a cell can be deduced from the intensity of the 90° light scatter. These parameters can be measured simultaneously with fluorescence from a surface marker or nuclear label; multiparameter data can then be used to generate three-dimensional histograms.

For the determination of ploidy in tumours by DNA analysis, which has prognostic value, the cell suspension has to be stained; propidium iodide and ethidium bromide are dyes most frequently used, but both bind also to RNA (this can be eliminated with RNAse treatment); acridine orange is also used and it has the remarkable property of fluorescing green when intercalated with double-stranded nucleic acids and orange-red when bound to single-stranded nucleic acids, thus enabling simultaneous analysis of DNA and RNA in the same sample. Single parameter analysis of DNA/cell

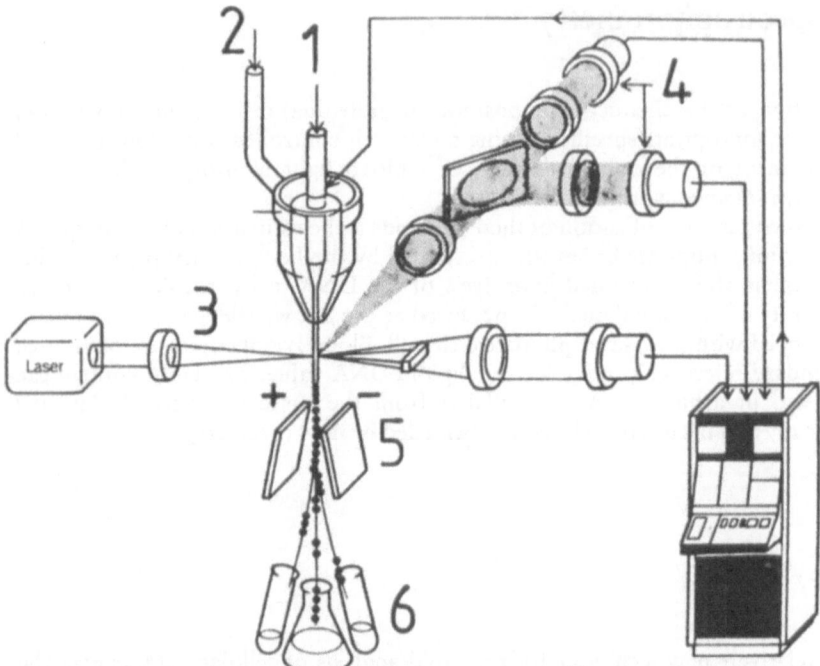

Fig. 10.3. Diagram of a flow cytometer with cell-sorting facility. Single-cell suspension (*1*) surrounded by sheath fluid (*2*) passes through the interrogating laser beam (*3*). The emitted fluorescent light (assuming that the cells have been labelled directly or indirectly with a fluorescent dye) strikes the photodetectors (*4*) and the signal is fed to the computer. A signal from the computer intermittently charges the fluid stream causing it to be deflected appropriately, according to preset conditions, as it passes between charged plates (*5*). The sorted cells are collected (*6*). (Courtesy of Becton Dickinson)

enables rapid determination of cell-cycle data in a cell population; cells in the Go/G1 phase will have the normal diploid quantity of DNA (i.e. 2C), cells in G2 and M phase will contain 4C DNA, and S-phase cells will occupy a position between the 2C and 4C peaks when the data are displayed graphically. Tetraploid or aneuploid populations can be detected and quantified by their position relative to the diploid population (Fig. 10.4). Most tumours contain infiltrating lymphocytes which can be used as an internal diploid standard, or the instrument can be calibrated with diploid cells from another source.

Immunofluorescent techniques can be used for the analysis of cells by surface marker expression. The cell suspension is pretreated with an antibody specific for the cell population that is to be detected and quantified; either the antibody itself bears the fluorochrome or the antibody is in turn detected by a second labelled anti-body in an indirect procedure (see Chap. 4 for an explanation of direct and indirect immunofluorescence).

Another facility on most machines is cell sorting. The fine stream of fluid carrying the cell suspension is gently vibrated to break it up into small droplets each carrying no more than one cell. The droplets then fall between two oppositely charged metal plates. If the detected parameters of each cell as it passes through the interrogating laser beam satisfy preset conditions, the instrument having first been programmed

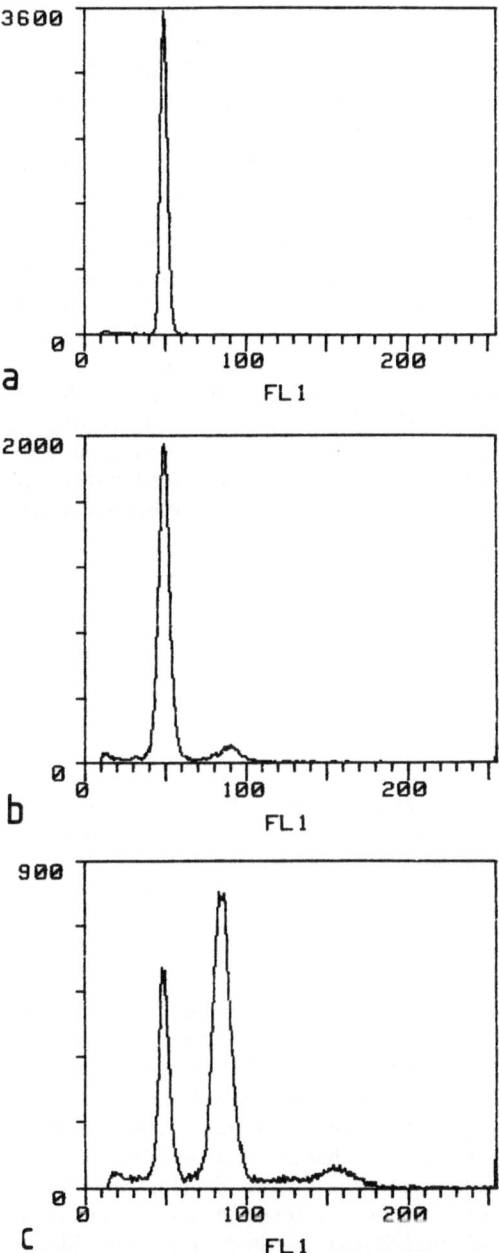

Fig. 10.4a–c. DNA profiles of carcinomas of the breast analysed by flow cytometry. The nuclear DNA of the cell suspension is stained by a fluorescent dye, and the fluorescent intensity and number of cells are plotted on the *horizontal* and *vertical axes* respectively. a Diploid standard derived from peripheral blood lymphocytes; the single peak is attributable to almost the entire cell population being in the G_1/G_0 phase of the cell cycle. b Diploid carcinoma with a G_2/M peak (maximum in flourescence intensity channel 90) denoting active DNA synthesis. c Tetraploid carcinoma in which the extreme right-hand peak is attributable to the tetraploid cells in the G_2/M phase of the cell cycle. The areas between the peaks are proportional to the number of cells in S phase

according to which cell population is to be sorted, a positive or negative charge is placed on the stream causing the droplets bearing the cells to be sorted to deviate from the main stream under the influence of electrostatic forces. The droplets deviated from the main stream are collected for further examination or study. For example, if one needed to separate out T-lymphocytes from a mixed cell population the suspension would first be treated with a pan-T-cell monoclonal antibody in either a direct or indirect immunofluorescent procedure so that all the T-lymphocytes as defined by the monoclonal antibody were now fluorescent. As each cell passes through the laser beam the emitted fluorescence is analysed and if it exceeds or falls within (i.e. "gated") preset threshold limits a signal is sent to charge the stream to coincide with the arrival of the droplet containing the fluorescent T-cell, which is then deviated by electrostatic attraction into the collecting tube. Cells can be analysed at the rate of up to 10 000 per second and sorted at a rate of up to 5000 per second.

The principal limitation of flow cytometry is the requirement for a single cell suspension which, for the analysis of solid tissues, has to be created by enzymatic dispersion. This process may damage some cells or selectively liberate an unrepresentative population. However, flow cytometric analyses of solid tissues processed in this way have been performed successfully, and it is even possible to analyse archival material stored as paraffin wax blocks (Hedley et al. 1983).

Practical Applications

The application of stereological and other quantitative aids to biopsy interpretation is the subject of active exploration. The potential or current value of quantification in histopathology is illustrated by the examples which follow.

Metabolic Bone Disease

Bone is an ideal candidate for quantitative analysis. Cancellous bone contains readily identifiable components—marrow, osteoid, mineralised bone—distributed almost at random. Changes in the relative proportions of osteoid and mineralised bone are often subtle, even in patients with clinically significant metabolic bone disease; the changes may be overlooked or overinterpreted without quantitative analysis. Also, bone alters naturally with age so that pathological changes may be difficult to distinguish from the normal process of senile involution. Further developments in the treatment of patients with Paget's disease, osteoporosis, osteomalacia, renal osteodystrophy, etc., demand accurate quantitative methods for bone biopsy assessment.

The relative volumetric proportions of marrow space, osteoid and mineralised bone can be quantified visually (e.g. by point-counting) or by semi-automatic or automatic image analysis. Other useful variables include the number of birefringent lamellae in osteoid seams and the number of osteoblasts and osteoclasts over a given area of trabecular surface. Technical details can be found elsewhere (Dunnill et al. 1967; Woods et al. 1968).

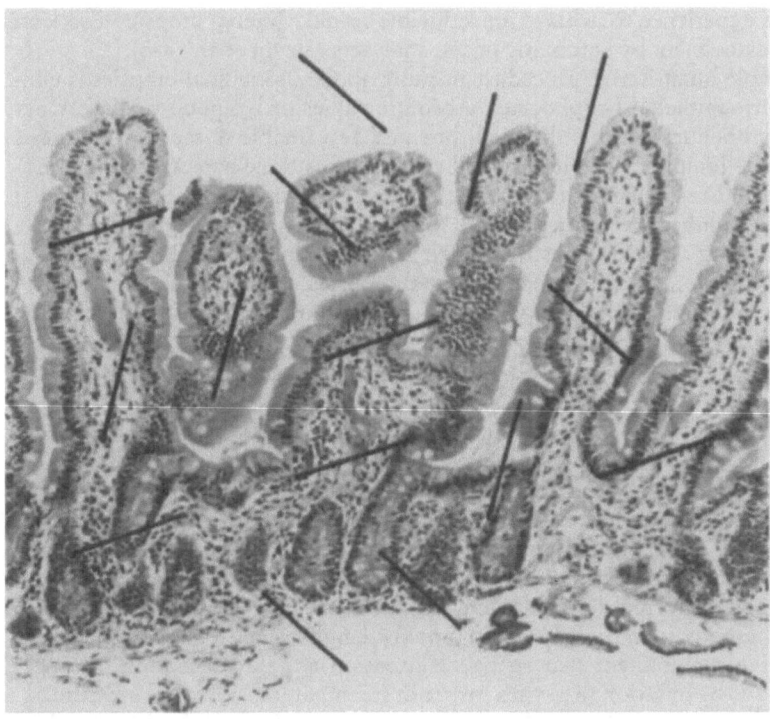

Fig. 10.5. Estimation of surface-to-volume ratio in a jejunal biopsy. The template devised by Weibel (1963) is superimposed on the section according to the method of Dunnill and Whitehead (1972). The method of calculation is given in the text.

Jejunal Biopsies

The simplest quantitative assessment of jejunal biopsies is the measurement of villous height, normally 4–5 times the basal zone thickness. Orientation is important for quantitative assessment of most anisotropic tissues, the villi and crypts should be sectioned longitudinally. Surface area is also a useful index and one of clinical import- ance because it partly determines the absorptive capacity of the mucosa. Mucosal biopsies are, however, prone to deformation and other artifacts so that simple linear measurement is often as invalid as would be assessment by purely visual means. To overcome this, Dunnill and Whitehead (1972) advocated the routine use of a simple method for determining the surface-to-volume ratio of the mucosa based on the grid devised by Weibel (1963), a series of 15 lines of equal length. The ends of each line are used as points for volumetric analysis of the mucosa by point-counting; intersec- tions between the lines and the mucosal surface provide an index of surface density (Fig. 10.5). The Weibel grid can be used either as an eyepiece graticule in a microscope or drawn on paper to receive a projected image. If C is the number of times the lines of the grid cut the mucosal surface, and H is the number of times the line-ends hit the mucosa, then $C:LH$ will give an indication of surface-to-volume ratio. The line length, L, is included to allow comparison with other biopsies that may be asses- sed with different sizes of Weibel templates. This rapid method is easy and needs relatively little experience to achieve reproducible results. Jejunal biopsies also lend themselves to assessment by automatic image analysers (Slavin et al. 1980).

Another simple quantitative procedure usefully applied to jejunal biopsies is enumeration of intraepithelial lymphocytes. Absolute values of lymphocyte density per unit area of epithelium can be obtained, but just as valuable is the less laborious estimation of the lymphocyte epithelial-cell ratio. This is done by counting lymphocyte and epithelial-cell nuclei; the normal range is 4–40 lymphocytes/100 epithelial cells. Excessive numbers of intraepithelial lymphocytes are seen in gluten-sensitive enteropathy, among other diseases. Although of little value in the differential diagnosis of malabsorption by assessment of jejunal biopsies, intraepithelial lymphocyte counts are quite useful in the comparison of repeat biopsies to assess therapeutic responses.

Respiratory Tract

Quantitative methods have been applied to the assessment of chronic bronchitis and emphysema. The relative depth of bronchial mucosa spanned by the mucous glands, the Reid index, is normally less than one-third and an important variable in the appraisal of chronic bronchitis (Reid 1960). Weibel's original work (Weibel 1963) on the stereology of normal lung parenchyma paved the way for the application of quantitative methods to the study of pulmonary emphysema. Vital to pulmonary stereology is the requirement that stringent attention is paid to fixation of lungs; this must be done by inflation (e.g. with formalin steam) at physiological pressure.

Stereological methods find little place in the interpretation of respiratory tract biopsies, but they are used for post-mortem specimens by those interested in emphysema.

Tumours

Aspects of neoplastic growth which have been studied include the stereological assessment of tumour structure and the quantification of cellular DNA in tumour-cell populations by microspectrophotometry or flow cytometry.

The volumetric composition of tumours has been determined by point-counting. For example, tumour cells account for only about 20% of the total volume of most scirrhous breast carcinomas; the remainder comprises stromal elements (Underwood 1972). Similar studies have been done in lung cancer (Gerstl et al. 1974). This information may be prognostically useful, but its significance awaits full evaluation. Morphometric analysis of nuclear profiles has been suggested as a useful method for characterising human tumour-cell populations (Stenkvist et al. 1978).

Microspectrophotometric methods, although now superseded by the more rapid technique of flow cytometry, have been used to obtain frequency distribution analyses of the DNA content of neoplastic cells in tumours (Fig. 10.6). The degree of aneuploidy or polyploidy may be used as a prognostic aid and as a guide to the selection of the most efficacious type of treatment (Dixon and Stead 1977). Obviously, whole cells rather than sectioned cells must be studied if all the DNA in the cell is to be assayed. Precautions are necessary to ensure that the sampling method (e.g. imprints, suspensions) yields a representative population of cells and that stromal contamination is minimised; with standard cytophotometry, however, the observer selects the cells for assay by direct microscopy.

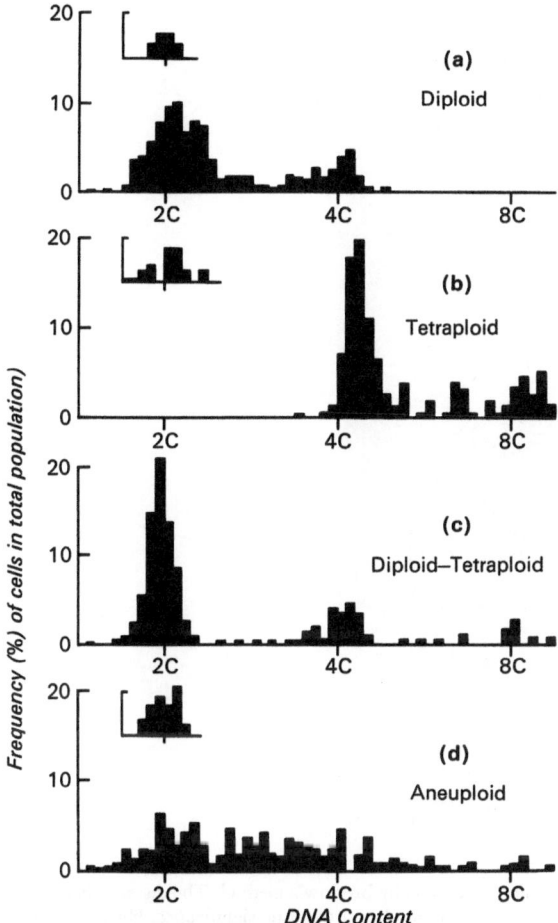

Fig. 10.6a–d. Characteristic DNA frequency distributions found in tissue imprints of a series of carcinomas of the cervix by Feulgen microdensitometry. The standard for the diploid (2C) state, shown as *inserts* in a, b, and d, was obtained from quiescent lymphoid cells infiltrating the tumours (Dixon and Stead 1977). The information can now be obtained more rapidly by flow cytometry (see Fig. 10.4).

Applications of flow cytometry to cytology and histopathology have been reviewed recently by Lovett et al. (1984). Ploidy measurements are under intensive investigation as possible prognostic indices in a wide variety of tumours. For example, the ploidy index of colorectal adenocarcinomas correlates with survival and appears to be independent of histological grade and surgical stage (Woolley et al. 1982). Flow cytometry assessment of ploidy resulted in the earlier detection of recurrent bladder carcinoma than would have been evident by routine cytology (Klein et al. 1982).

Mitotic activity is regarded as an important criterion for the diagnosis of malignancy in some connective tissue neoplasms such as those of smooth muscle origin. Commonly, the histopathologist seeks guidance from a monograph or tumour atlas about the frequency of mitotic figures in the lesion in question; the frequency of mitotic figures seen in the malignant variant of a particular tumour is usually

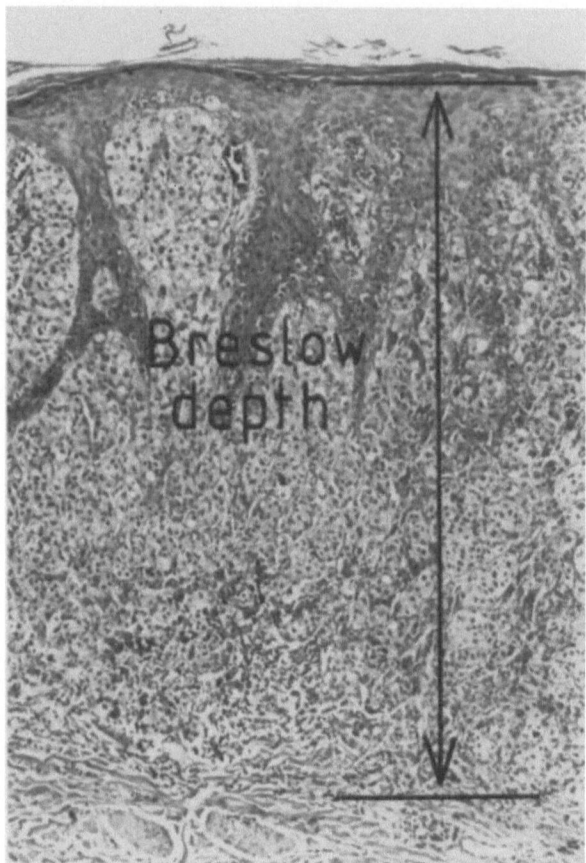

Fig. 10.7. Measurement of the depth of a nodular melanoma by Breslow's method. The distance between the stratum granulosum and the deep edge of the melanoma has prognostic significance. Haematoxylin and eosin, × 107

expressed as a certain number per high-power microscope field. This procedure is prone to two uncertainties. First, the apparent frequency of mitotic figures in a high-power microscope field will depend on the size of the high-power field (this will differ with different microscopes and their objective lenses) and the thickness of the tissue section (usually approximately 5μm but subject to inevitable variation for technical reasons). Ellis and Whitehead (1981) estimate that, for the aforementioned reasons, mitosis counting may be subject to a 600% variation between different institutions. And then there is the question of observer error in the reproducibility of mitosis counts. Silverberg (1976) submitted ten smooth muscle tumours of the uterus to six pathologists and asked them to count the number of mitoses per ten high-power fields; there was considerable variation and the panel of patholologists were unanimous in only four of the ten tumours in assigning them to the malignant category if mitotic activity was the only criterion. It seems that pyknotic nuclei may have been mistaken for mitotic figures in some instances. Despite these cautions, mitotic counting remains one of the more useful and simpler quantitative methods that can be used to assist in the distinction between benign and malignant connective tissue tumours.

The measurement of tumour depth is an accurate guide to prognosis in malignant melanomas (Breslow 1970). It is necessary to measure the distance between the stratum granulosum and the deep edge of the melanoma (Fig. 10.7). This can be done with a calibrated graticule in the focal plane of the eyepiece of the microscope. A simpler way is to use the vernier scale on the stage: the stage carrying the slide is moved so that the stratum granulosum coincides with the edge of the field and the vernier scale reading is recorded; a second reading is taken with the deep edge of the tumour at the edge of the field; the difference between the two vernier readings gives the depth of the tumour.

References

Baak JPA, Kurver PHJ, Boon ME (1982) Computer-aided application of quantitative microscopy in diagnostic pathology. Pathol Annu 2: 287–306

Bohm N, Sandritter W (1975) DNA in human tumours: a cytophotometric study. Curr Top Pathol 60: 151–219

Breslow A (1970) Thickness, cross-sectional areas and depth of invasion in the prognosis of cutaneous melanoma. Ann Surg 172: 902–908

Chalkley HW (1943) Methods for the quantitative morphologic analysis of tissues. J Natl Cancer Inst 4: 47–53

Chalkley HW, Cornfield J, Park H (1949) A method for estimating volume-surface ratios. Science 110: 295–297

Dixon B, Stead RH (1977) Feulgen microdensitometry and analysis of S-phase cells in cervical tumour biopsies. J Clin Pathol 30: 907–913

Dunnill MS (1968) Quantitative methods in histology. In: Dyke SC (ed) Recent advances in clinical pathology, series V. Churchill, London, pp 401–416

Dunnill MS, Whitehead R (1972) A method for the quantitation of small intestinal biopsy specimens. J Clin Pathol 25: 243–246

Dunnill MS, Anderson JA, Whitehead R (1967) Quantitative histological studies on age changes in bone. J Pathol Bacteriol 94: 275–291

Ellis PE, Whitehead R (1981) Mitosis counting: a need for reappraisal. Hum Pathol 12: 3–4

Gerstl B, Switzer P, Yesner RA (1974) A morphometric study of pulmonary cancer. Cancer Res 34: 248–254

Hedley DW, Friedlander ML, Taylor IW, Rugg CA, Musgrove EA (1983) Method for analysis of cellular DNA content of paraffin-embedded pathological material using flow cytometry. J Histochem Cytochem 31: 1333–1335

Kamentsky LA, Melamed MR, Derman H (1965) Spectrophotometer: new instrument for ultrarapid cell analysis. Science 150: 630–631

Klein FA, Herr HW, Sogani PC, Whitmore WFJr, Melamed MR (1982) Detection and follow-up of carcinoma of the urinary bladder by flow cytometry. Cancer 50: 389–395

Lovett EJ, Schnitzer B, Keren DF, Flint A, Hudson JL, McClatchey KD (1984) Application of flow cytometry to diagnostic pathology. Lab Invest 50: 115–140

Koenig SH, Brown RD, Kamentsky LA, Sedlis A, Melamed MR (1966) Efficacy of a rapid cell spectrophotometer in screening for cervical cancer. Cancer 21: 1019–1026

Mayhew TM, Cruz Orive LM (1974) Caveat on the use of the Delesse principle of areal analysis for estimating component volume densities. J Microsc 102: 195–207

Reid L (1960) Measurement of the bronchial mucous gland layer: a diagnostic yardstick in chronic bronchitis. Thorax 15: 132–141

Silverberg SG (1976) Reproducibility of the mitosis count in the histologic diagnosis of smooth muscle tumours of the uterus. Hum Pathol 7: 451–454

Slavin G, Sowter C, Robertson K, McDermott S, Paton K (1980) Measurement in jejunal biopsies by computer-aided microscopy. J Clin Pathol 33: 254–261

Stenkvist B, Westman-Naeser S, Holmquist J, Nordin B, Bengtsson E, Vegelius J, Eriksson O, Fox CH (1978) Computerized nuclear morphometry as an objective method for characterizing human cancer cell populations. Cancer Res 38: 4688–4697

Underwood JCE (1972) A morphometric analysis of human breast carcinoma. Br J Cancer 26: 234–237

Van Dilla MA, Trujillo TT, Mullaney PF, Coulter JR (1969) Cell microfluorimetry: a method for rapid fluorescent measurements. Science 163: 1213–1214

Weibel ER (1963) Principles and methods for the morphometric study of the lung and other organs. Lab Invest 12: 131–155

Weibel ER (1969) Stereological principles for morphometry in electron microscope cytology. Int Rev Cytol 26: 235–302

Weibel ER, Elias H (1967) Introduction to stereologic principles. In: Weibel ER, Elias H (eds) Symposium on quantitative methods in morphology. Springer, Berlin Heidelberg New York, pp 89–98

Weibel ER, Kistler GS, Scherle WR (1966) Practical stereologic methods for morphometric cytology. J Cell Biol 30: 23–38

Woods CG, Morgan DB, Paterson CR, Gossman HH (1968) Measurement of osteoid in bone biopsy. J Pathol Bacteriol 95: 441–447

Woolley RC, Schreiber K, Koss LG, Karas M, Sherman M (1982) DNA distribution in human colonic carcinomas and its relationship to clinical behaviour. J Natl Cancer Inst 69: 15–22

Reporting and Classification of Biopsy Diagnoses

The essence of the biopsy report is *communication*. It is the medium through which the pathologist routinely communicates to the physician or surgeon. It is important to ensure that it is not the weakest link in the biopsy procedure; there would be little point in taking the biopsy, safely transmitting it to the laboratory, having sections cut by expert histotechnologists, and making an accurate diagnosis, if the final opinion on the biopsy was eventually lost in an incomprehensible mass of tangled verbiage.

The Biopsy Report

The precise style of a biopsy report is a personal matter for the individual pathologist and will not be considered. Nor will the detailed structure of reports be discussed because this is a matter for individual laboratories and the sort of specimens they deal with. Most reports, however, consist of two or more sections.

Gross Description

This is usually dictated when the specimen or biopsy is "cut up". This description may have subsequently to be edited when the histological appearances are seen in order to avoid ambiguity or inconsistency, without necessarily changing the accuracy of the description. The specimen and any grossly detectable lesions should be properly documented with weights and linear measurements where appropriate. The description should leave the precise nature of any lesions completely open; deductions should be avoided. For example, it would be wrong to describe a lymph node that was eventually reported as showing sarcoidosis as "A lymph node 3 × 2 × 1 cm. The cut surface shows uniform replacement by white tumour tissue". It would be better to describe the cut surface as "uniformly white with a firm consistency". The gross description should therefore be objective rather than deductive. Allowance must be made, however, for artifactual changes induced in the colour and consistency of tissues by fixation. It would be misleading to describe a lymph node fixed in Bouin's fluid as yellow when the colour has been imparted to the node by immersion in a fixative containing picric acid. Ideally, specimens should be described when they are fresh; blocks can be taken either later when the tissue is fixed or immediately from the fresh tissue and trimmed as necessary after fixation.

Histology

The histological description in a report should substantiate the final opinion that the report conveys. The alert physician or surgeon often places more weight on a well-founded opinion that on an opinion unsupported by written evidence.

Detailed histological descriptions are superfluous for many routine specimens or common lesions where there is little or no clinical doubt about the diagnosis (e.g. chronic cholecystitis, acute appendicitis), but certainly any liver, renal, lung, or intestinal biopsy, just to select a few examples, should be described in some detail. However, not everything that one sees in a histological section merits inclusion on the final report; mention of *Demodex folliculorum*, a mite, in skin biopsies from the face or nipple (common sites of infestation) is unnecessary and would worry anyone but a dermatologist. The report should not be dominated by lengthy descriptions of conspicuous or exotic features which may have little or no diagnostic importance. Also, negative statements are often just as important as positive statements and tell the reader of the report what abnormalities have been excluded as well as those that have been found.

The integrity of the tissue and adequacy of the sample should be commented on without fear of offending the person who took the biopsy. It would be wrong to state baldly that an endoscopic biopsy showed no evidence of any tumour if the specimen consisted of a solitary minute crushed tissue fragment.

Other Investigations

Most diagnoses are based on the interpretation of gross appearances and routine histology. Sometimes further work is necessary, such as electron microscopy or immunohistology, which may delay the report especially if the material has to be sent elsewhere. It is usually acceptable to issue a preliminary report containing a tentative diagnosis subject to confirmation by other investigations. In cases of appreciable doubt it is far better to discuss the problem and render a verbal opinion as to the differential diagnosis. Credibility is all too easily demolished by having to rescind an initial diagnosis even if it was well founded on the evidence available at that time. The same applies to obtaining a second opinion, where it is invariably better to await the expert's view before the definitive written report is issued for insertion in the patient's case notes.

Summary

Most reports contain a summary of the final diagnosis either as a separate statement or included in the main body of the report. This should be a clear statement on which the clinician can take therapeutic or further investigative action. If previous biopsies from the same patient are available it is useful not only to report the current biopsy but also to compare it with those previously seen. If the biopsy has been interpreted without knowledge of the full clinical picture, this should be stated where it may be pertinent.

The diagnosis may be qualified by the use of phrases like "consistent with" or "suggestive of", unavoidable in cases of doubt. If there is doubt, this must be stated. Sometimes the clinician who sees a biopsy report in which it is stated that "the appearances are suggestive of" some entity thinks that the pathologist is being evasive. If the merely suggestive nature of the appearances is supported by a description of the lesion this should be sufficient to allay the clinician's suspicions. Diagnostic doubt may arise because the biopsy has been taken only from the vicinity of a lesion rather than from the lesion itself. This point can be emphasised by stating that "there is no unequivocal evidence of any tumour in this particular sample". A few millimetres can make the difference between a certain diagnosis or no diagnosis at all. "Pathologists have not yet learned to make tumour diagnoses from tissue taken near the lesion. Extrasensory perception has not yet become a part of the orthodox surgical pathologist's armamentarium" (Foraker 1959).

A histological report is incomplete and relatively useless if the pathologist fails to answer the questions that clinicians naturally ask about specific problems; the pathologist must adopt their point of view and consider what therapeutic actions they are choosing between. For example in reporting a thyroid carcinoma, while "papillary" means an *appearance* to pathologists, to surgeons it means a certain type of *behaviour* and *prognosis*. Surgeons do not need to be bothered with the presence of follicular areas in a papillary thyroid carcinoma; their only concern is that the lesion is going to behave like a papillary tumour and should be treated as such. Reporting a rare lesion will naturally raise questions from the clinician who will want to know what its likely behaviour and response to treatment is; these questions can be pre-empted by an informative report.

Reporting Tumours

The assessment of tumour biopsies and resections constitutes a major part of most pathologists' workload and therefore, and for reasons related to the nature of the disease, it warrants special consideration. For most tumours, three aspects need to be considered and reported on:

— Tumour type or classification (e.g. benign or malignant, histogenesis)
— Tumour grade (e.g. well, moderately or poorly differentiated)
— Tumour stage (e.g. Dukes' stage A, B, or C for colorectal carcinomas).

Tumour resections should be examined and sampled so that the tumour class (or type), grade and stage can be deduced from the gross pathological description and the histology.

A relatively small biopsy may suffice for classification, but will be less satisfactory for grading because the degree of differentiation seen in the biopsy may not be representative of the lesion as a whole and it may not contain the least well differentiated area which is likely to have the greatest clinical impact in terms of determining prognosis. Staging is, of course, impossible with a small biopsy, though metastases revealed in small biopsies of organs other than that which contains the primary neoplasm may contribute information from which the tumour stage can be determined.

With regard to the reporting of the degree of differentiation of a tumour, this should be assessed in the least well differentiated area; this is likely to represent the most aggressive cell population in the tumour and will often dictate the most appropriate treatment. The best differentiated area guides the pathologist to the type of tumour, the least differentiated area guides its treatment.

The written biopsy diagnosis of "carcinoma-in-situ" should be used with caution. If indeed the lesion really does show features of carcinoma-in-situ, there are even some who would dispute the verifiability of that diagnosis (Smith 1978). "Severe dysplasia" may have similar biological implications and the possible advantage that these words do not carry the same penalty for insurance purposes. The unqualified statement that there is "carcinoma-in-situ" in a small biopsy may also tempt the conservative clinician to "wait and see" while the invasive part of the lesion, not included in the biopsy, progresses.

The expression "carcinoma-in-situ" should be used with special caution in the gastrointestinal tract. It is, first, ambiguous in that it could mean either carcinoma within pre-existing glandular spaces or carcinoma confined to the mucosa, and, second, the limitations of superficial endoscopic biopsies do not often adequately exclude the presence of invasive carcinoma in the immediate vicinity. The clinical significance of intramucosal carcinoma depends on the site. Because lymphatics are not present in *colorectal* mucosa above the muscularis mucosae, superficial carcinomas confined to this zone have no scope for lymphatic metastasis and should not be reported as "carcinoma". However, lymphatics do occur in *gastric* mucosa so there is greater potential for metastasis by superficial or "early" carcinomas of the stomach (Morson and Dawson 1979).

"Atypical hyperplasia" can be regarded as a terminological variant of dysplasia. The term is commonly used for borderline endometrial proliferations, but Azzopardi (1979) has cogently argued against the use of words like "dysplasia" and "atypical hyperplasia" when reporting breast lesions; the surgeon should be told unequivocally whether the lesion, in the pathologist's opinion, is benign or malignant to avoid unnecessary mastectomies.

Disease Classifications

Not all disease classifications are histopathological, but the histopathologist is often the best individual to classify those diseases that are diagnosed on biopsy. The main objectives in classifications are to facilitate communications and to satisfy the need for uniformity for epidemiological purposes. Some may feel, however, that standardised classifications become stagnant, inhibit free thinking, and retard further assessment of nosological problems. These are valid arguments and should motivate periodic revision of most existing classifications.

"Lumping" and "Splitting"

Those who use and devise classifications have themselves been classified as either "lumpers" or "splitters" (Bohrod 1971). The "lumpers" use coarse classifications composed of broad categories whereas the "splitters" use fine classifications with a large number of narrowly defined categories (Chap. 1). "Splitters" are epitomised by the dermatologists whose system has been described as "a colourful cornucopia of concatenating cutaneous classification" (Feinstein 1967).

The advent of modern nosology dates back to seventeenth century England and the work of Thomas Sydenham (1624–1689). Disease classification had been based on symptoms (e.g. fever, consumption) since the time of Hippocrates (ca. 460–ca. 377 B.C.). Sydenham's innovative classification was still clinical rather than morbid anatomical, but he defined disease categories by recognising either groups of clinical manifestations that often occurred together in the same patient (a cluster of symptoms and signs, or a syndrome) or a consistent pattern in the way the condition evolved (its temporal course or natural history). The trend that he started was extended by the revolution of morbid anatomy in the nineteenth century and continues to be developed through the application of modern technological methods.

Nomenclature and Codes

Various standard systems have been devised to satisfy the overwhelming need for an internationally acceptable nomenclature of disease that would permit comparability studies on the incidence of disease in different locations. These essentially numerical systems or codes are based on standard nomenclature and are not really classifications in themselves. They must depend, however, on uniform classifications if the encoded data is going to be of value to those engaged in epidemiological studies or those attempting to retrieve individual cases. The World Health Organization (WHO) has been largely responsible for devising standard classifications for tumours (Sobin 1977) and various other diseases, those of the liver for example (Anthony et al. 1978). Other agencies have also encouraged the use of standard schemes (Fogarty International Center Proceedings 1976).

Early standard classifications, notably that devised in 1893 by Jacques Bertillon (1851–1922), were confined to mortality statistics. Bertillon's classification did, however, become the standard in many countries and was recommended for use by the American Public Health Association in 1898. After Bertillon's death the Health Organization of the League of Nations undertook further revisions of his classification. In 1946 responsibility passed to the World Health Organization and a revised classification combining mortality and morbidity was produced two years later. This "International Classification of Disease" (ICD) has undergone several further revisions under the continuing auspices of the WHO (World Health Organization 1977).

In the 1950s the College of American Pathologists started work on a multiaxis coding system (topography, morphology, aetiology and function) that was eventually published in 1965 as the "Systematized Nomenclature of Pathology" (SNOP). This was adopted by many pathologists in various countries. It became clear that the SNOP coding system could be usefully applied to medicine in general and in 1976 the College of American Pathologists published a "Systematized Nomenclature of Medicine" (SNOMED), fully compatible with SNOP and a logical extension of it.

For oncology, an area in which epidemiological studies drawing on standard nomenclature and codes are of paramount importance, the WHO in collaboration with the International Agency for Research on Cancer and the National Cancer Institute in the United States developed the "International Classification of Diseases for Oncology" (ICD–O) (World Health Organization 1976). The coding system was based on MOTNAC ("Manual of Tumour Nomenclature and Coding" published by the American Cancer Society in 1968) which had in turn been based on ICD and SNOP.

As a further refinement the WHO has produced the "International Classification of Procedures in Medicine" (World Health Organization 1978), an innovation adopted for trial by the 29th World Health Assembly in 1976. Biopsy and excisional procedures are included, with distinctions drawn between simple biopsies (endoscopic, punch, percutaneous needle biopsies and smears) coded 1–4 .. where the last two digits are site specific, and surgical biopsies (incisional, open operation or excisional, but excluding therapeutic excision) coded 1–5 .. where the last two digits are again site specific. Excisional operations are coded 5 – ...

Problems in Tumour Classification and Grading

The classification and nomenclature of tumours remains an outstanding problem for those seeking epidemiological data. For example, renal cell carcinoma, adenocarcinoma of the kidney, Grawitz tumour, hypernephroma and clear-cell carcinoma are all used by different pathologists to describe the same histogenetic type of tumour (Sobin 1977). Sobin (1972) reviewed 57 papers on lung cancer in which a total of 13 distinct classification systems had been used. These are problems of nomenclature or terminology. Another problem is to establish a set of criteria which the individual pathologist can use to assign a particular tumour reliably to its correct place in an ordained classification.

The WHO has embarked on a programme for the production of the "International Histological Classification of Tumours" (Sobin 1971). For each tumour site a panel of experts draws up a standard nomenclature, classification, and set of definitions which is then evaluated in participating laboratories. The classification is discussed and amended before publication by the WHO as a "Blue Book" (e.g. Kreyberg et al. 1981) in which the recommended nomenclature is given. The final set of 25 volumes is reviewed periodically. Numerical coding can be obtained from ICD–O, SNOMED, or SNOP for each tumour separately indexed in the WHO classifications.

Unfortunately, not everyone is satisfied with the WHO classifications; this is reflected in the continued proliferation of published classifications even within a single organ system. This may be based on conceptual differences about the histogenesis of specific tumours or on results from the application of novel technology to the investigation of certain tumour systems.

Lymphoma classifications, of which there are many, illustrate well some of the difficulties in classifying tumours and the clinical use to which this information is put. In the 1960s lymphomas were classified on purely morphological evidence (e.g. 'poorly-differentiated lymphocytic' in Rappaport's classification). In the mid-1970s several new classifications appeared that reinterpreted lymphoma morphology in the light of new knowledge about the relationship of lymphocytes to the cells of the germinal centres of lymphoid organs (e.g. 'follicle cells, predominantly small' in the British National Lymphoma Investigation classification, 'malignant lymphoma, centrocytic' in the Kiel classification, and 'small cleaved follicle-centre cell lymphoma' in the classification of Lukes and Collins). In the following decade the availability of monoclonal antibodies to T-cell surface markers and their application to immunohistological studies of lymphomas showed that a significant number of lymphomas that would previously have been regarded as of B-cell type were, in fact, T-cell tumours—at least as judged by their immunophenotype. Now it is possible to perform genotypic analyses on lymphomas and look for the gene rearrangements that characterize T- and B-lymphocytes; this is revealing some discrepancies between immunophenotype and genotype (e.g. cells bearing surface immunoglobulin and therefore assumed to be of B-cell lineage may unexpectedly show gene rearrangement typical of cells of T-cell lineage). What then is to be the 'gold standard' for lymphoma classification— morphology, immunophenotype, or genotype? And yet despite all the effort that has gone into devising these classifications and the assigning of individual lesions to categories within them, often all that the medical oncologist needs to know is "is it high-grade or low-grade lymphoma?", such is the limited availability of therapy as specific as the pathologists' classifications.

Grading of tumours is a separate issue from their histogenetic classification. At a crude level we speak of a tumour as being "well differentiated", "poorly differentiated" or "anaplastic". The distinction between each grade is often ill defined and to some extent arbitrary (Chap. 5). More rigid criteria for tumour grading have been proposed such as those of Broders (1932) originally for squamous-cell carcinoma of the lip, Kernohan and Sayre (1952) for astrocytomas, and Bloom and Richardson (1957) for breast carcinoma. Grading often has prognostic and therapeutic implications, but just as important may be stromal features such as lymphoreticular infiltration (Underwood 1974) and the local extent of the tumour either through direct invasion or by permeation of vascular channels and lymph node metastases (e.g. Dukes' staging of colorectal carcinoma). Not all grading systems are equally easy to use; significant differences within and between observers have also been found (Stenkvist et al. 1979). Heterogeneity within a single tumour often presents a problem, in which case the general rule is to score the lesion on its least well-differentiated areas; these are likely to contain the most biologically aggressive neoplastic tissue.

The TNM (tumour topography, lymph node status, distant metastasis) system of tumour staging was originated by the International Union Against Cancer (IUCC) for general application to tumour staging irrespective of primary site; the numerical stage is derived from the variable TNM scores. Each field in the TNM staging system is further qualified by a number or letter suffix: suffix 'o' denotes no evidence of tumour; numbers 1 to 4 denote progressive increases in tumour size or organ involvement; 'X' denotes insufficient evidence for assessment of status; 'Tis' denotes carcinoma in situ. Further suffixes can be added if greater precision is required: for breast carcinoma, for example, 'T1biii' denotes a primary tumour fixed to pectoral muscle or fascia (b), between 1 and 2 cm in diameter (iii). For details of TNM system and its current applications the reader is directed to the recent review by Henson (1985).

Data Storage and Retrieval

One of the objects of encoding diagnoses is to make it easier to store and retrieve lesions for retrospective analyses. Even in a small laboratory this facility is a considerable asset for purely internal purposes and, if an internationally accepted code such as SNOP is used, it allows outside agencies easier access to the local files. The encoded data may be stored in a classified file consisting of either a separate copy of the full biopsy report or a series of cards identified by separate codes, each card bearing the patients' name or laboratory numbers of the lesions given that particular code number (Kennedy 1977). An alternative system is to programme a computer to code diagnoses written as short standard statements (e.g. "breast, fibroadenoma"—topography given before morphology as in the sequence for SNOP coding) which are then stored with the specimen identification codes as the biopsy report is typed (Coles and Slavin 1976).

Any coding system is prone to error. Enlander (1975) reports that, on average, 76% of lesions could be retrieved when coded manually by residents at Stanford University Medical Center. Retrieval must be done intelligently and with a wide ambit to avoid overlooking cases that have been miscoded, perhaps through personal idiosyncrasies rather than error. Ulcerative colitis, for example, (SNOP topography code 67– for colon) might be SNOP-coded as either 67–4103 for acute ulcerative colitis, 67–4303 for chronic ulcerative colitis, or 67–4003 for ulceration. Individual laboratories should try to use consistent codes for lesions in which there is potential for variable coding. Lesions miscoded through random errors would of course be less easy to retrieve.

Clinical Assimilation

The main purpose of taking a biopsy is to seek a valid appraisal of its nature with a view to obtaining a diagnosis or other therapeutically useful information. The biopsy alone may provide the diagnosis or it may have to be considered in conjunction with other data (e.g. biochemical, clinical) and thus only partly contribute to the eventual diagnosis.

The final link in the biopsy chain is the assimilation of the report by those directly involved in the care of the patient. They may have to weigh the pathologist's opinion against other factors to resolve a possibly conflicting clinical picture. The biopsy alone may actually make the diagnosis or, if the evidence is less clear-cut, merely suggest it or reinforce a clinical opinion. Certain lesions may be excluded within the limitations of the biopsy or sampling. In many instances the written opinion has to be qualified by direct discussion; there is then more freedom to express intuitive opinions or "hunches".

The most useful biopsy report from the clinician's point of view is one that conveys a totally unqualified diagnosis (e.g. in summary "Well-differentiated invasive rectal adenocarcinoma"). Less useful is the qualified diagnosis (e.g. in summary "Appearances are very strongly suggestive of malignancy"). Sometimes the diagnosis given may be only partly qualified (e.g. in summary "This is a poorly differentiated malignant neoplasm, possibly adenocarcinoma"). Obviously if the biopsy contains a pathognomonic feature that is exclusive to a single disease then the clinician should be entitled to an unqualified diagnosis. Qualified diagnoses arise when the biopsy contains some of the features of a disease but lacks the pathognomonic sine qua non (e.g. a lymph node containing caseating granulomas but no histologically detectable acid-fast bacilli). Often an unequivocal biopsy diagnosis can be made only when a certain number of criteria are fulfilled, no single feature being regarded as absolutely pathognomonic (e.g. mucin depletion, crypt abscesses, ulceration in ulcerative colitis). Sometimes it is possible for the biopsy to exclude certain diseases unequivocally, especially those that produce diffuse changes in tissues from which adequate biopsies are obtained (e.g. an architecturally normal liver biopsy devoid of stainable iron excludes untreated haemochromatosis), without necessarily making a positive diagnosis.

Clinicians may find difficulty in appreciating how confident the pathologist feels about the diagnosis. This is particularly so in neoplastic diseases where the institution of treatment invariably depends upon obtaining a tissue diagnosis; any doubt about the diagnosis or precise classification of a tumour may have important therapeutic implications. Russell (1980) has proposed a code that could be appended to the biopsy report to indicate roughly the confidence level of the diagnosis; C1 denotes a confidence level > 98%, C2 75%–98%, C3 50%–75%, and C4 < 50%. A tentative diagnosis of cancer attracting a confidence level of C4 would receive less clinical emphasis; further biopsies or other investigations might have to be undertaken before the diagnosis of cancer could be made with acceptable certainty in that patient.

Biopsies are often misleadingly spoken of by clinicians as "showing" a particular lesion, as though the diagnosis was manifestly obvious. What is really meant is that the biopsy was "interpreted as showing" a given lesion, but abbreviation to "showing" gives a false impression of incontestability. It carries the implication that there is no scope for questioning the diagnosis; it may not be considered worthwhile to institute a review of the biopsy if fresh information about the patient comes to light.

The biopsy report is therefore an exercise in communication, a means of conveying an opinion as to the biopsy diagnosis to those who will take therapeutic action on it. Eloquent biopsy reports are no substitute for good rapport between histopathologist and clinician.

References

Anthony PP, Ishak KG, Nayak NC, Poulsen HE, Scheuer PJ, Sobin LH (1978) The morphology of cirrhosis: recommendations on definition, nomenclature, and classification by a working group sponsored by the World Health Organization. J Clin Pathol 31: 395–414

Azzopardi JG (1979) Problems in breast pathology. Saunders, Philadelphia, Toronto, p 167

Bloom HJG, Richardson WW (1957) Histologic grading and prognosis in breast cancer. A study of 1409 cases of which 359 have been followed for 15 years. Br J Cancer 11: 359–377

Bohrod MG (1971) What is a pathologic diagnosis? Pathol Annu 6: 197–208

Broders AC (1932) Practical points on the microscopic grading of carcinoma. NY State J Med 32: 667–671

Coles EC, Slavin G (1976) An evaluation of automatic coding of surgical pathology reports. J Clin Pathol 29: 621–625

College of American Pathologists (1965) Systematized nomenclature of pathology. College of American Pathologists, Chicago

College of American Pathologists (1976) Systematized nomenclature of medicine, College of American Pathologists, Skokie

Enlander D (1975) Computer data processing of medical diagnoses in pathology. Am J Clin Pathol 63: 538–544

Feinstein AR (1967) Clinical judgement. Williams and Wilkins, Baltimore, p 85

Fogarty International Center (1976) Disease of the liver and biliary tract. Standardization of nomenclature, diagnostic criteria and diagnostic methodology. DHEW (NIH), Washington (proceedings, no. 22)

Foraker AV (1959) On the importance of "weasel words" in surgical pathology. Surg Gynecol Obstet 109: 637–638

Henson DE (1985) Staging for cancer: new developments and importance to pathology. Arch Pathol Lab Med 109: 13–16

Kennedy A (1977) Basic techniques in diagnostic histopathology. Churchill Livingstone, Edinburgh London New York, pp 12–20

Kernohan JW, Sayre GP (1952) Tumors of the central nervous system. Armed Forces Institute of Pathology, Washington DC, pp 22–42

Kreyberg L, et al. (1981) Histological typing of lung tumours, 2nd edn. WHO, Geneva

Morson BC, Dawson IMP (1979) Gastrointestinal pathology. Blackwell, Oxford London Edinburgh Melbourne, p 609

Russell WO (1980) The pathologic diagnosis of cancer: a crescendo of importance in current and future therapies. Am J Clin Pathol 73: 3–11

Smith C (1978) Carcinoma in situ. Hum Pathol 9: 373–374

Sobin LH (1971) The World Health Organization's programme for the histopathological definition and classification of tumours. Methods Inf Med 10: 120–122

Sobin LH (1972) Multiplicity of lung tumour classifications. In: Grundmann E, Tulinius H (eds) Current problems in the epidemiology of cancer and lymphomas. Springer, Berlin Heidelberg New York, pp 30–35

Sobin LH (1977) Standardization and the histopathology of tumours. Histopathology 1: 87–92

Stenkvist B, Westman-Naeser S, Vegelius J, Holmquist J, Nordin B, Bengtsson E, Eriksson O (1979) Analysis of reproducibility of subjective grading systems for breast carcinoma. J Clin Pathol 32: 979–985

Underwood JCE (1974) Lymphoreticular infiltration in human tumours: prognostic and biological implications. Br J Cancer 30: 538–548

World Health Organization (1976) International classification of diseases for oncology. WHO, Geneva

World Health Organization (1977) Manual of the international statistical classification of diseases, injuries and causes of death. WHO, Geneva

World Health Organization (1978) International classification of procedures in medicine, vol. 1. WHO, Geneva

12   Quality Assessment and Control

Histopathologists are "more than machines into which one can feed coloured slides at one end and expect quality controlled answers to pour out at the other" (Wright 1975), but at the same time one must seek to avoid a situation in which "pathological audit might be based on the frequency of irate clinicians looking for a better service" (Dudley 1975).

The now well-recognised fallibility of diagnostic histopathologists has led to the development of quality assessment and audit schemes. Some of these schemes have operated nationally though on a voluntary basis. Others are forms of professional accreditation, such as the examination for membership of the Royal College of Pathologists (London, U.K.), without which the individual cannot proceed beyond a certain point in their chosen career.

How the quality of a histopathological diagnosis can be assessed and the magnitude of diagnostic fallibility in histopathology are discussed in this chapter.

Sources of Diagnostic Error

In diagnostic histopathology and cytopathology errors can occur at the stages of:

— Sampling the patient
— Sampling the biopsy or resection
— Misreading or misinformation about the clinical context
— Technical processing
— Perception of the abnormalities present
— Interpretation of the abnormalities perceived
— Communication of the diagnosis
— Transcription of the written diagnosis

Most quality assessment schemes examine the error chain from the point at which abnormalities are or should be perceived, though many departments will often audit their performance of all stages of specimen handling and diagnosis as and when concern about quality and efficiency arises.

It is useful, therefore, if some sort of external control can be applied. This is done in many places on a voluntary basis. National and international meetings with slide seminars are now commonplace and give histopathologists an opportunity to compare their opinions with those of accredited experts. An alternative approach is to seek a second opinion on difficult or ambiguous cases. Willingness to seek a second opinion is a commendable asset and should never be regarded as an indication of incompetence. At the same time it is essential to strive for the right compromise between the pathologist who "does" 10 000 per annum and never requests an outside opinion, and the pathologist who immediately refers all problems without thoroughly investigating them first. In actual practice it is better to send unstained sections so that the expert is not distracted by local staining peculiarities. All relevant clinical information should be submitted; most experts find it helpful to know the referring pathologist's opinion so that the problem can be rapidly identified and discussed in reply (Sissons 1978).

The "Correct" Diagnosis

Various formal quality assessment schemes have been proposed. In such schemes the diagnosis proffered by each participant would be judged against the standard or "correct" diagnosis. This "correct" diagnosis can be obtained in three ways.

1. The "correct" diagnosis is that given by a majority of participants. The reliability of this method would depend upon the expertise of the group. Of the three methods, this is probably the least reliable.
2. The "correct" diagnosis is that given by an acknowledged doyen. This is acceptable if the views of the expert are justified by previous experience.
3. The "correct" diagnosis is that which is verified by observing the behaviour of the lesion in a patient, its response to specific treatment, or the unequivocal demonstration of a specific aetiological agent.

In the context of quality control, the distinction between discrete and continuous diagnostic categories is important (Langley 1978). With discrete categories there is theoretically only one possible "correct" diagnosis (e.g. invasive squamous-cell carcinoma of skin; aspergillus pneumonia; gluten-sensitive enteropathy). Continuous categories, exemplified by the dysplasia-carcinoma spectrum of cervical neoplasia, are not amenable to the concept of a "correct" diagnosis, the lesion can only be assigned a place within the spectrum. A problem peculiar to continuous categories is observer drift, a measurable change in the interpretation of a standard set of biopsies by a single observer over a period of time. In addition to the dysplasia-carcinoma spectrum, drift may also be encountered in tumour grading according to differentiation. A high degree of consensus and consistency can be developed by a group of histopathologists who work in the same centre; new recruits learn the consensus and trained personnel drift to conform with the diagnostic ambience of the majority. The danger is that the group may be lulled into a sense of self-righteous confidence, while they deviate more and more from others working elsewhere whose opinions may have greater veracity.

Evaluation of Diagnostic Methods

Quite separate from the evaluation of the subjective aspects of making a histopathological diagnosis is the assessment of the technical aspects of a diagnostic procedure. This may, for example, examine the quality of histological sections and their staining. Such a scheme was tried in Wales for 3 years (Barr 1978) and has since been extended to laboratories in England with the support of the Royal College of Pathologists and the Institute of Medical Laboratory Sciences (Barr and Williams 1982).

The accuracy of any diagnostic procedure, including technical and interpretative aspects, can be assessed in terms of its specificity and sensitivity:

$$\text{specificity} = \frac{\text{true negative}}{\text{true negative} + \text{false positive/suspicious}}$$

$$\text{sensitivitity} = \frac{\text{true positive}}{\text{true positive} + \text{false negative}}$$

In the context of cancer, "malignant" and "benign" or "non-neoplastic" would be substituted for "positive" and "negative" respectively.

Sensitivity and specificity constitute the accuracy of a diagnostic method. The precision of a diagnostic method is the extent to which it can assign an unknown lesion to the narrowest possible diagnostic category. For example, a method may readily and accurately diagnose malignancy but not so frequently allow a precise classification of the lesion according to tumour type (Payne et al. 1979).

Evaluation of Performance

In assessing the histological diagnosis it is necessary to evaluate three potential types of error—repeatability, validity and bias (Silcocks 1983). Repeatability may be assessed within-observer (i.e. intra-observer—the same histopathologist examining the same slide on two or more occasions without knowledge of the opinion expressed on previous occasions) or between-observer (i.e. inter-observer—when two or more histopathologists examine the same slide under similar conditions). Validity is the extent to which a histopathologist's diagnosis coincides with the "correct" diagnosis determined by any of the criteria summarised on p. 188. Bias is the extent to which a histopathologist's diagnoses systematically deviate from the "correct" diagnoses on a series of biopsies.

If the maximum amount of useful information is to be obtained from a quality assessment scheme it is essential that the gathered data is analysed to yield information about repeatability and bias. Silcocks (1983), reviewing this topic, recommends the use of Cohen's kappa statistic; this is a form of correlation coefficient in which o represents chance agreement and +1 represents perfect agreement, and has been applied to the assessment of liver biopsies (Theodossi et al. 1980), melanomas (Holman et al. 1982), and the grading of rectal carcinomas (Thomas et al. 1983).

Illustrative Examples

Virtually all assessment schemes run by some external agency are open to the criticism that they are not conducted under the same conditions that operate during the normal working day in a diagnostic laboratory. The test sections are easily spotted as being part of a special exercise and the participant is therefore alerted to the fact that competence and expertise is about to be tested. Ideally, the test slides should be insinuated among the rest of the day's work, but there are obvious practical difficulties in so doing.

Most histological diagnoses have to be computed from the assessment of a series of separate features, each contributing to the overall appearance, and it can be difficult to dissect out the feature(s) responsible for observer disagreement. This has been done with mitosis counting in smooth muscle tumours of the uterus (Silverberg 1976); this is commonly regarded as an important criterion for the diagnosis of malignancy in these lesions. Sections of ten smooth muscle tumours were examined by six pathologists who were asked to count the number of mitoses per ten high-power fields (a procedure which in itself is subject to non-observer variables—see pp. 173–174). There was considerable variation in mitosis counts, with unanimity in only four of the ten tumours; this raised serious doubts about the validity of using mitotic counts as the sole criterion for malignancy in these lesions.

The construction of quality assessment schemes and the extent of the diagnostic fallibility of histopathologists is illustrated by the following examples (see also pp. 10–12).

Liver Biopsy Interpretation

In a detailed analysis of the performance of 6 pathologists Theodossi et al. (1980) reported that histological features associated with a high kappa value (i.e. strong agreement) include fatty change, siderosis and Mallory's hyalin, and features associated with a low kappa value (i.e. no significant agreement) include venous congestion, bile duct destruction and granulomas. Full agreement between six histopathologists on the final diagnosis of 60 liver biopsies occurred in only nine instances, all of which were examples of alcoholic liver disease.

Bladder Tumour Grading

Fifty-seven bladder tumours were graded according to the World Health Organization (W.H.O.) system by six histopathologists whose experience and expertise ranged from one who had been head of the Department of Pathology at the Free University Hospital in Amsterdam for 27 years and was a member of the Amsterdam Bladder Tumour Registry to another who had only 4 years' experience in histopathology (Ooms et al. 1983). The tumours were assessed in two sessions separated by 7 months. The pathologists sat together in the same room but were unable to confer, except for two pathologists who were asked simply to rank the tumours according to the estimated degree of malignancy. It was thus possible to compare intra- and inter-observer variability. Intra-observer consistency was remarkably low; about half of the tumours were graded differently between the two sessions, differing by more than one grade in nearly 8% of cases. Correlation coefficients for inter-observer grading was ranged from +0.46 to +0.67. The authors concluded that the inconsistencies were "disturbingly high".

Breast Carcinoma Grading

Stenkvist et al. (1979) examined the inter- and intra-observer reproducibility of the W.H.O. (Scarff and Torloni 1968), Black and Speer (1957), and Hartveit (1971) grading systems applied to 179 breast carcinomas. Each tumour was evaluated on two different occasions, separated by 6 months, by two histopathologists. The authors concluded that all three grading systems had low inter- and intra-observer reproducibility.

Six histopathologists independently graded 158 breast carcinomas by the W.H.O. method (Scarff and Torloni 1968). Five of the histopathologists had trained simultaneously in the same institution and the sixth was one of their tutors. In only 23 tumours did all the histopathologists agree on the grading, 20 of which were grade 1 or 3. Forty-one tumours were classified under all three grades. The authors conclude that, since agreement was confined predominantly to grade 1 or 3 tumours and these could be readily assigned to well and poorly differentiated categories respectively, there is no purpose in using the W.H.O. grading system because it is insufficiently reproducible (Delides et al. 1982).

National and Institutional Schemes

Royal College of Pathologists of Australasia. An Australian survey provided detailed information on the degrees of difficulty or error posed by particular lesions, and compared the performances of trainee and trained pathologists (Royal College of Pathologists of Australia 1969, 1970, 1971, cited by Penner 1973). Correct diagnosis rates ranged from 92.7% for Arias-Stella reaction, 95.8% for Whipple's disease and 93.9% for dermatopathic lymphadenitis, to 37.6% for tuberculous endometritis, 11.0% for alveolar rhabdomyosarcoma and 7.2% for peritonitis arenosa. For the 12 lesions in each set, the ratio of correct diagnoses offered by fully trained specialists as compared with trainees was—9.6 :6.8 (1969), 9.0 :8.0 (1970), and 8.0 :5.5 (1971). General pathologists tended to occupy an intermediate position. The "correct" diagnoses in the Australian survey were submitted by acknowledged experts.

More recently in response to pressure to introduce accreditation of laboratories, the Royal College of Pathologists of Australasia has introduced a more formal method of quality control in histopathology (Cooke 1984). The objective was to examine three aspects : the diagnostic skills of the participating pathologists; their work habits; and the technical proficiency of the laboratory staff. With regard to diagnostic accuracy, the "correctness" of the submitted diagnoses was assessed by comparing them with the majority opinion. The circulated cases covered a wide spectrum of organs and systems and included both common lesions and rarities. The results of this ambitious survey are detailed in the article by Cooke (1984) to which the reader is referred.

College of American Pathologists. An Interlaboratory Histologic Evaluation scheme, under the auspices of the College of American Pathologists, was employed by Macartney et al. (1981). It was based on the assumption that the relative incidence of diseases and disease subtypes is similar in comparable institutions. For example, in all institutions with a comparable patient population in terms of race, geographic location and demography, the ratio of the frequency of two diagnoses should be very similar; types of analysis included the ratio of invasive to intraduct breast cancer, age versus site scattergrams for a particular lesion (e.g. squamous carcinoma of skin), relative frequency of entities within a continuous spectrum of disease (e.g. grades of dysplasia). Comparing the ratios of diagnosed invasive:intraduct breast cancers in five hospitals in the Birmingham, U.K. area, the authors reported a variation of −39% to +141% around the mean ratio. The ratio of in-situ :invasive cervical cancers ranged from 0.4 to 0.9 in four hospitals in the U.K. and from 1.05 to 3.85 in nine hospitals in the U.S.A.; this reveals differences between countries and between institutions within the same country.

An earlier survey under the auspices of the College of American Pathologists also used routine workload material (Penner 1973). Each participant submitted the first breast lesion, lymph node, skin tumour, etc. received after a certain date. Slides were then circulated to all participants without clinical data. The "correct" diagnosis was that reached by a majority of the participants. Each diagnosis was recorded as a SNOP number (Systemized Nomenclature of Pathology). Analysis of the results showed the greatest degree of consensus with epithelial skin tumours and the least with lymph nodes. Pathologists concurred on average, with their own original diagnoses in 78% of the sections. However, clinical data were not given in the survey.

St. Thomas's Hospital. Quality evaluation within a single department was the subject of a British survey (Owen and Tighe 1975). Random sections were submitted to participants with the patient's age and sex and the site of the lesion. The final consensus was regarded as the "correct" diagnosis. Senior histopathologists (possessing the Membership of the Royal College of Pathologists) differed on average from the consensus diagnosis on 6.8% of sections, compared with a rate of 19.0% for junior colleagues. A distinction was drawn between discrepancies that were genuinely misleading and might result in inappropriate treatment, and those which were of a trivial nature. It was proposed that this type of study could be done to assess an individual's proficiency with a view to the delegation of responsibility for routine biopsy reporting.

Self Assessment

The American Society of Clinical Pathologists has devised self assessment tests for all branches of pathology including surgical pathology. For surgical pathology the test consists of a set of multiple-choice questions linked to histology slides or electron micrographs and are devised to test diagnostic competence as well as pathological knowledge. Scores are reported back to the participant in a form that indicates how the participant's performance compares with others who have taken the same test. Participation is voluntary and confidentiality is guaranteed (Flynn 1980).

Assessment of Errors of Procedure and Communication

Most audit schemes examine the diagnostic act. Few schemes have been devised to audit the antecedent events (describing the specimen and taking blocks for histology) and the final step of typing and signing out the report.

Even a relatively simple procedure, like harvesting lymph nodes from the mesentery of a resected colorectal cancer, is subject to considerable variation. Twenty-two histopathology departments participated in a colorectal cancer project comprising 2046 specimens. The lymph node harvest from the whole population was 5.25 ± 4.7 (mean $\pm$ S.D.); however, the harvest from colorectal resections examined in individual hospitals ranged from 1.0 ± 1.6 to 11.2 ± 5.8, a difference significant at the $p < 0.001$ level, and in one hospital no nodes were recovered from 54.2% of resections (Blenkinsopp et al. 1981).

Murthy and Derman (1981) have reported their experience of this type of audit at the Riverside Methodist Hospital at Columbus, Ohio, U.S.A.; they examined transcription errors, adequacy of gross descriptions, use of terminology and delays in reporting. Each of the participating pathologists took it in turn to review the surgical pathology reports. Typographic and transcription errors were "frequent but not of a serious nature, except, on occasion, when the word 'no' had been omitted in a diagnosis such as 'no evidence of carcinoma is seen'". Other typographical errors included the substitution of "crotic cartilage" for "eroded cartilage" and the transposition of a microscopic diagnosis of "segments of fallopian tube" to a specimen of a vocal cord polyp!

Conclusions

Quality assessment surveys reinforce our knowledge of the fallibility of histopathological diagnoses and serve to identify lesions which tend to be erroneously interpreted. Most histopathologists would unhesitatingly participate, others may feel that it is a means of proficiency testing with possible implications for accreditation. Coercion should not be necessary if the objectives of the schemes are clearly presented to potential participants. The end result should be presented as a measure of fallibility, not culpability.

References

Barr WT (1978) Technical quality control in histopathology. J Clin Pathol 31: 996–998

Barr WT, Williams ED (1982) Value of external quality assessment of the technical aspects of histopathology. J Clin Pathol 35: 1050–1056

Black MM, Speer FD (1957) Nuclear structure in cancer tissues. Surg Gynecol Obstet 105: 97–102

Blenkinsopp WK, Stewart-Brown S, Blesovsky L, Kearney G, Fielding LP (1981) Histopathology reporting in large bowel cancer. J Clin Pathol 34: 509–513

Cooke RA (1984) Quality control in anatomic pathology: experience in Australia and New Zealand. Pathol Ann [Part 1]: 221–248

Delides GS, Garas G, Georgouli G, Jiortziotis D, Lecca J, Liva T, Elomenoglou J (1982) Intralaboratory variations in the grading of breast carcinoma. Arch Pathol Lab Med 106: 126–128

Dudley HAF (1975) Audit and the pathologist. Proc Roy Soc Med 68: 634–637

Flynn FV (1980) Competence to practise and self-assessment. Bull Roy Coll Pathol 32: 16–18

Hartveit F (1971) Prognostic typing in breast cancer. Br Med J 4: 253–257

Holman CDJ, James IR, Heenan PJ, Matz LR, Blackwell JB, Kelsall GRH, Singh A, Ten Seldam REJ (1982) An improved method of analysis of observer variation between pathologists. Histopathology 6: 581–589

Langley FA (1978) Quality control in histopathology and diagnostic cytology. Histopathology 2: 3–18

Macartney JC, Henson DE, Codling BW (1981) Quality assurance in anatomic pathology. Am J Clin Pathol 75 [Suppl]: 467–475

Murthy MSN, Derman H (1981) Quality assurance in surgical pathology: personal and peer assessment. Am J Clin Pathol 75 [Suppl]: 462–466

Ooms EC, Anderson WA, Alons CL, Boon ME, Veldhuizen RW (1983) Analysis of the performance of pathologists in the grading of bladder tumours. Hum Pathol 14: 144–150

Owen DA, Tighe JR (1975) Quality evaluation in histopathology. Br Med J i: 149–150

Payne CR, Stovin PG, Barker V, McVittie S, Stark JE (1979) Diagnostic accuracy of cytology and biopsy in primary bronchial carcinoma. Thorax 34: 294–299

Penner DW (1973) Quality control and quality evaluation in histopathology and cytology. Pathol Ann 8: 1–19

Scarff RW, Torloni H (1968) Histological typing of breast tumours. (International histological classification of tumours, No. 2). WHO, Geneva

Silcocks PBS (1983) Measuring repeatability and validity of histological diagnosis—a brief review with some practical examples. J Clin Pathol 36: 1269–1275

Silverberg SG (1976) Reproducibility of the mitosis count in the histologic diagnosis of smooth muscle tumours of the uterus. Hum Pathol 7: 451–454

Sissons HA (1978) On seeking a second opinion. J Clin Pathol 31: 1121–1124

Stenkvist B, Westman-Naeser S, Vegelius J, Holmquist J, Nordin B, Bengtsson E, Eriksson O (1979) Analysis of reproducibility of grading systems for breast carcinoma. J Clin Pathol 32: 979–985

Theodossi A, Skene AM, Portmann B, Knill-Jones RP, Patrick RS, Tate RA, Kealey W, Jarvis KJ, O'Brian DJ, Williams R (1980) Observer variation in assessment of liver biopsies including analysis by kappa statistics. Gastroenterology 79: 232–241

Thomas GDH, Dixon MF, Smeeton NC, Williams NS (1983) Observer variation in the histological grading of rectal carcinoma. J Clin Pathol 36: 385–391

Wright EA (1975) Quality control in histopathology. Proc R Soc Med 68: 619–622

13   The Autopsy

Autopsy and biopsy work are inseparable activities; the same technical and interpretative skills are used for both. The autopsy diagnosis obviously has less immediate clinical impact than does the biopsy diagnosis, but the contribution of an efficient autopsy service to the overall standard of clinical care in any hospital is surely undeniable. However, there has recently been much questioning of the role and status of the autopsy now that so many sophisticated techniques are used to investigate and diagnose disease during life. There is no doubt that there is currently less interest in the autopsy than there was in the nineteenth century, during the age when discoveries in morbid anatomy often heralded advances in clinical practice.

The borderline between diagnostic histopathology and autopsy work is artificial and purely administrative. This is amply illustrated by the post-mortem examination of aborted fetuses in surgical pathology laboratories (Berry 1980), an examination which should involve careful external examination, photography and radiology where necessary, and histology (Anonymous 1981).

History of the Autopsy

King and Meehan (1973) comprehensively reviewed the history of the autopsy and its impact on clinical practice.

The early dominance of animism—that worldly events are controlled by invisible supernatural forces—held back the development of the autopsy, and indeed medical progress, because this philosophy denied the possibility that disease might be due to internal disturbances of bodily structure and function. Nevertheless, there is one aspect of animistic philosophy that bears on the development of the autopsy. Reference has already been made (Chap. 1) to haruspicy, an example of divination in which the future might be foretold by examining animals' entrails, especially the liver (i.e. hepatoscopy) considered to be the seat of the soul. This ancient practice can be traced back to c. 3500 B.C. The future could be predicted by the diviner, or "haruspex" in the Roman era, who carefully noted and interpreted the changes in hepatic morphology. Thus, our ancestors learnt much about anatomy and certain abnormalities through haruspicy and also, of course, from the slaughtering and preparation of animal flesh for food (Sigerist 1967).

Naturalism, on the other hand, ascribed disease to natural causes, though its early exponents such as Hippocrates (468–377 B.C.) placed greater emphasis on the humors—the real and hypothetical body fluids—than on the solid structures such as internal organs that might be investigated by autopsy. Medical historians have concluded that human dissections were not performed by the Greeks before the third century B.C., but at that time it was established practice in Alexandria to investigate human anatomy in health and disease. Erasistratus (c. 310–250 B.C.) noted for example that the liver was hard in deaths from dropsy but soft in cases of snake bite. Human dissections in Alexandria probably continued to the time of Galen (130–200 A.D.) (Edelstein 1967).

Human dissections and autopsies disappear from the historical record until the Middle Ages when first reference to these activities in England and Italy can be found. The discovery of hitherto unknown abnormalities of internal structures in deaths from a variety of causes assured the future of the autopsy in the practice of medicine.

Religious attitudes almost certainly retarded the autopsy and anatomical studies. For example, although Christian churches have never actually prohibited these activities it was not until the fifteenth and sixteenth centuries that the Catholic church gave its formal approval. The sudden death of Pope Alexander in 1410 was investigated by autopsy, and calculi in the kidneys, bladder and gall bladder were found at an autopsy in 1556 on Ignatius Loyola. In 1533 an autopsy was done on conjoined twins in what is now the Dominican Republic; both twins had been baptised, after much thought, but there was still doubt as to whether there was one or two souls. Autopsy disclosed two complete sets of internal organs and the priest was reportedly content that he had done the right thing (Chavarria and Shipley 1924). Jewish tradition strongly respects the human body and any form of dissection was strictly forbidden, an attitude maintained by orthodox Jews until this century when a law permitting strictly controlled autopsies was passed by the Knesset (Kottler 1957).

The widespread use of autopsies was also resisted by popular attitudes such as were recorded in Germany in 1670 (Jarcho 1971) and in the Republic of Lucca in 1699 (Castiglioni 1947).

These difficulties were sufficiently overcome to enable some quite large series to be collected and published, resulting in an impressive compilation of over 3000 cases by Theophilus Bonetus (1620–1689) in his "Sepulchretum", first published in 1679 and based on the experiences of numerous authors from Galen onwards. The genius of Giovanni Battista Morgagni (1682–1771) found its expression in his detailed meticulous records of morbid anatomical changes disclosed at autopsy; his approach was totally objective and having made his observations he then sought to correlate them with the clinical features of the illness. This scientific approach was further extended by Xavier Bichat (1771–1802), who allegedly performed about 600 autopsies in his final year.

Although Rokitansky (1804–1878) allegedly performed about 30,000 autopsies, he didn't use microscopy until his later years. Although his contributions to morbid anatomy remain outstanding and probably without parallel since Morgagni, his conclusions and interpretations were necessarily limited in comparison with Virchow (1821–1902) who readily accepted the microscope (Rother 1966).

By the end of the nineteenth century the autopsy had become an established part of the contemporary medical scene. William Osler was pre-eminent among contemporary practitioners for his extensive experience of autopsies in Montreal and Philadelphia and his verdict has had a lasting influence on the role of the autopsy:

"To investigate the causes of death, to examine carefully the condition of organs, after such changes have gone on in them as to render existence impossible and to apply knowledge to the prevention and treatment of disease, is one of the highest objects of the physician . . ." (William Osler, cited by McPhee and Bottles 1985).

The current status of the autopsy and some special considerations about techniques are reviewed in the following sections.

The Autopsy and Clinical Audit

Although the status of the autopsy in providing material for medical education remains unchallenged, the declining autopsy rates in most hospitals reflect a waning of clinical interest in the procedure as a means of verifying diagnoses made during life. Critics have argued that the application of sophisticated modern diagnostic tests during life has refined the accuracy of clinical diagnoses, so that little more can be gleaned from the examination of lifeless, often autolysed, organs should the patient eventually succumb. But many major studies have affirmed the status of the autopsy in clinical audit (Goldman 1984).

Cabot (1912) studied 3000 autopsies and reported that important disorders were often missed clinically: 77% of cases of spinal tuberculosis; 61% of hepatic cirrhosis; 26% of lobar pneumonia; etc. He believed these errors resulted from the limitations of diagnostic methods rather than from clinical misjudgement. In 1923, Wells reported that 33% of tumours discovered at autopsies in two Chicago hospitals had not been diagnosed during life; tumours of lung and liver were most frequently underdiagnosed (Wells 1923). These findings had much impact and assured high autopsy rates in institutions wishing to achieve the highest standards of medical education and practice.

After an interval of about 30 years, a further series of publications appeared reaffirming the frequency of missed diagnoses disclosed by autopsy. Gruver and Freis (1957) reviewed errors found in 1106 autopsies between 1947 and 1953 at the Veterans Administration Hospital, Washington, D.C. Bacterial pneumonia, meningitis, endocarditis and tumours accounted for half of the diagnoses missed clinically; 6% of the 1106 cases had been misdiagnosed. Contributory factors were: failure to obtain a routine screening test on admission; failure to follow-up a sign or symptom previously noted; failure to follow-up a previous abnormal laboratory result; and failure to review preconceived diagnoses. Goldman et al. (1983) noted that, in 1970, 12% of autopsies disclosed errors that if recognised and treated during life might have averted death or otherwise affected the outcome. The Edinburgh experience reported by Cameron et al. (1980) has been similar: the principal diagnosis was missed in 15% of 152 autopsies; in 5% the error was so significant that the patient might have benefited from appropriate treatment.

Role in Medical Education

Pathologists need no convincing that the autopsy has an important role in medical education. Indeed, recognition of the importance of autopsy pathology was probably a major influence on their choice of speciality. For most students the autopsy is the only opportunity they have of correlating their clinical assessment of an individual patient with the morbid anatomy; the diastolic murmur heard in life can be vividly correlated with the stenotic mitral valve seen at autopsy.

Some students have been alert to the problem of the declining interest in the autopsy and have taken steps to reverse the trend. At the June 1983 meeting of the American Medical Association the enlightened Medical Students Section sponsored Resolution 137 which has now become official AMA policy (Lundberg 1983):

"Resolved, That the American Medical Association develop and distribute to students and residents appropriate material stressing the importance and necessity of the autopsy as a service both to the medical community and to society as a whole; etc."

The autopsy may also have a role in the teaching of anatomy:

"If I were to start afresh to plan the teaching of anatomy, I certainly would not start with the cadaver. If we wanted a child to learn what a plum is we would not start by telling him to dissect a dried up old prune! At autopsy the tissues are still vivid." (Cope 1968).

Students who participated in an experimental programme at Stanford University in February 1973 were impressed by the "bright colours of the viscera, the moistness, the glistening appearance and softness of tissues and the flowing of blood when vessels were cut" (Castleman et al. 1974).

Hazards in the Autopsy Room

The principal hazards in the autopsy room are microbiological. Every case poses some risk, if only from the bacterial flora of the bowel, but some are especially hazardous and any local guidelines for dealing with specific problems must of course be rigorously followed. Wearing impervious gloves and an apron when performing an autopsy is routine in most countries and institutions. Clean working practices and regular washing and disinfection of all surfaces coming into contact with body fluids is to be expected.

Each institution will have its own guidelines for the conduct of autopsies, but national guidelines or requirements may exist for dealing with the special hazards such as hepatitis virus B and HIV carriers. In the United Kingdom such guidance is to be found in the "Code of Practice for the Prevention of Infection in Clinical Laboratories and Post-mortem Rooms" (HMSO 1978). The likely yield of useful information from an autopsy on risk cases must be weighed against the risk itself and all staff concerned with the autopsy must be informed that risk exists. Autopsies on high risk cases must never be done to satisfy idle curiosity; there must be an expectation that new information will be obtained.

The pathologist can be protected from infectious risks if the examination is conducted with the body inside a clear flexible plastic tent; negative pressure is maintained with a small pump and the dissection is carried out with the pathologist's arms within plastic invaginations in the tent wall (Trexler and Gilmour 1983).

Consent

Medicolegal requirements demand that the pathologist must identify the body; consider the possibility of the case falling under the jurisdiction of the coroner, medical examiner, or equivalent local officer; ascertain that there is written authorisation for the autopsy and note and heed any restrictions imposed by those who gave authorisation; and determine whether or not there is a known risk of infection that might render an autopsy inadvisable or restricted.

Considerable international or interstate differences exist concerning the legal authority to permit an autopsy (Schmidt 1983). In some countries an autopsy may be conducted 24 h after death so long as no objection has been made within that period; this enables clinical staff and pathologists to select cases for autopsy without actively seeking consent. In other places, such as the United Kingdom, the law requires consent to be actively sought. If religious beliefs prohibit an autopsy this can in certain situations be respected, though the public interest in cases of medicolegal importance may be considered to over-ride personal wishes. In California, an autopsy is prohibited if the deceased is known to have been a member of a religion, church or denomination relying solely on prayer for healing (Schmidt 1983).

General Techniques

While personal variations in autopsy techniques may exist, the end result should be the same: a clear demonstration of the ultimate and antecedent causes of death or illness and an adequate search for unexpected or incidental features.

The principal methods have been summarised by Ludwig (1972) and are named according to their provenance:

1. Virchow's method. Organs removed individually in the following order—brain, spinal cord, thoracic, cervical, abdominal.
2. Rokitansky's method. In situ dissection, combined with en bloc removal.
3. Ghon's method. Thoracic and cervical organs, abdominal organs, and, separately, urogenital organs are removed as blocks for subsequent dissection.
4. Letulle's method. Removal of cervical, thoracic, abdominal and pelvic organs in continuity as one block.

Of these methods, Ghon's with personal modification is probably the most widely used. Lettulle's method yields a heavy and rather cumbersome block of organs, but is the method of choice when there are known or suspected abnormalities immediately above or below or traversing the diaphragm. Lettule's method may also be preferred for fetuses and infants, the organ block being more manageable in these cases, but the reader is referred to Langley's detailed account of the perinatal autopsy for more comprehensive guidance (Langley 1971).

Special Methods

The routine autopsy consists of external examination, internal examination, and light microscopy as time and resources permit and the need indicates. Sometimes the circumstances of the case dictate a departure from routine practice and the use of special investigative methods. These are now summarised.

Microbiology

Bacteria, fungi or viral inclusions may be evident on light microscopy of autopsy tissues, but proper identification requires microbiological techniques.

For the diagnosis of septicaemia it is best to obtain a blood sample with the minimum risk of contamination. Perhaps the simplest expedient is a splenic stab: the abdomen is opened and the splenic capsule cauterised with a hot blade or iron; a sterile swab can then be inserted through the now sterilised splenic capsule into the pulp.

Sepsis in isolated organs, however, means that microbiological sampling must be targeted accordingly. For example, if a pulmonary infection is suspected, an appropriate sample of lung must be taken with precautions to minimize the risk of contamination by extraneous organisms.

A direct smear stained by Gram's method is useful to indicate the predominant organism because the importance of contaminating bacteria may be exaggerated by their overgrowth of subsequent culture (Ludwig 1972).

Post-mortem Biochemistry

Except in the forensic arena, where post-mortem toxicology is essential, biochemical methods have been rarely exploited in autopsy work.

Trump and his colleagues in Baltimore have been innovative in this area (Trump et al. 1975). They gathered a considerable amount of biochemical and other data through the "Immediate Autopsy Program" at the University of Maryland. Cases studied include those who were deemed irreversibly brain dead (according to the criteria developed by the Ad Hoc Harvard Committee—see JAMA (1968) 205: 337) or those who died with refractory shock. Legal authority to perform an immediate autopsy was provided by either the Medical Examiner's Office, in cases of violent or unattended natural death, or the Anatomical Gift Act of the State of Maryland. The immediate autopsy, commenced within minutes of death, involved removal of tissue from 19 sites, some of which was processed for mitochondrial analysis, histochemistry and other biochemical procedures. Following this, a conventional autopsy was performed. By 1975, 33 cases had been examined in this way and much correlative ultrastructural and biochemical information obtained.

More conventional perhaps is the biochemical analysis of body fluids from autopsies reviewed by Coe (1974). Blood is an obvious choice for autopsy biochemistry: it is reasonably accessible, available in relatively large volumes, data exists on normal ranges, and the values obtained can be compared with any obtained during life. But the biochemical analysis of post-mortem blood samples can give erratic results because of haemolysis or, possibly, bacterial growth. Cerebrospinal fluid is "cleaner", but relatively inaccessible particularly if large volumes are needed free of contaminating blood. Vitreous humor is probably the fluid of choice. It is less liable to be contaminated by the products of autolysis and as much as 2 ml of clear fluid can be aspirated from each eye.

The analysis of glucose illustrates some of the problems. Blood in peripheral vessels undergoes glycolysis at the rate of approximately 12.8 mg/dl/h, thus giving falsely low post-mortem glucose concentrations. Right atrial blood, however, contains abnormally high glucose concentrations due to glycogenolysis in the liver (Hamilton-Paterson and Johnson 1940; Hill 1941). Unexpectedly high concentrations have been found in carbon monoxide poisoning, raised intracranial pressure and respiratory obstruction.

Urea concentrations in blood seem to be remarkably stable in cadavers, though the apparent elevation compared with ante-mortem concentrations may be attributable to an agonal rise. Creatinine concentrations appear equally stable, but do not seem to show the agonal rise exhibited by urea; post-mortem concentrations closely resemble ante-mortem levels.

Enzyme concentrations in blood rise markedly after death. Acid phosphatase levels, for example, are over 20 times the normal ante-mortem value by 48 h after death (Enticknap 1960).

Serum electrolytes are extremely difficult to evaluate after death. Sodium concentrations decrease, the rate varying between individual cadavers. Potassium concentrations rise rapidly and so much that no meaningful conclusions can be drawn.

Although vitreous humor has the advantages of clarity, relative accessibility and freedom from contamination, it does have the disadvantage that no normal values exist with which to compare post-mortem values. However, it has been found that normal vitreous humor contains roughly half the serum concentration of glucose.

Vitreous humor can be readily aspirated by puncturing the sclera at the lateral canthus using a 20-gauge needle attached to a syringe (Leahy and Farber 1967).

Vitreous values for enzymes show no apparent correlations with disease or ante-mortem serum concentrations.

Electrolyte concentrations in vitreous humor appear relatively stable, sodium and chloride especially so. Potassium concentrations rise with so much individual variation as to render the analysis of limited value (Coe 1969). The vitreous is also useful for toxicology (Coe 1974).

Blood samples taken at autopsy for alcohol analysis, in cases where this may be medicolegal importance, must be taken into tubes containing inhibitors of fungal and bacterial growth otherwise falsely high values may be obtained (Vuori et al. 1983).

The Needle Autopsy

The use of needle techniques to remove tissue samples from the living body is well established clinical practice. Less well known is the use of needles to remove tissue from cadavers. The indications and advantages are:

1. Reducing the hazard of infection when dealing with high risk cases.
2. Removal of fresh tissue soon after death for detailed study (e.g. electron microscopy), having first obtained consent and fulfilled any other medicolegal requirements.
3. Where consent for a full autopsy has been refused, but relatives are willing to give consent to a less disfiguring needle sampling of the cadaver.

The major disadvantage is sampling error: the technique is unsuitable for detecting focal lesions (e.g. scanty small metastatic deposits in liver), and for sampling small organs (e.g. parathyroid glands). With skill and practice it is possible to obtain cores of liver, lung, kidney, muscle, spleen and prostate for routine histology with special stains, electron microscopy and immunohistochemistry (Underwood, Slater and Parsons 1983) (Fig. 13.1). Fluids and effusions can also be aspirated. Terry (1955), credited with the first description of this autopsy technique, suggested that it was

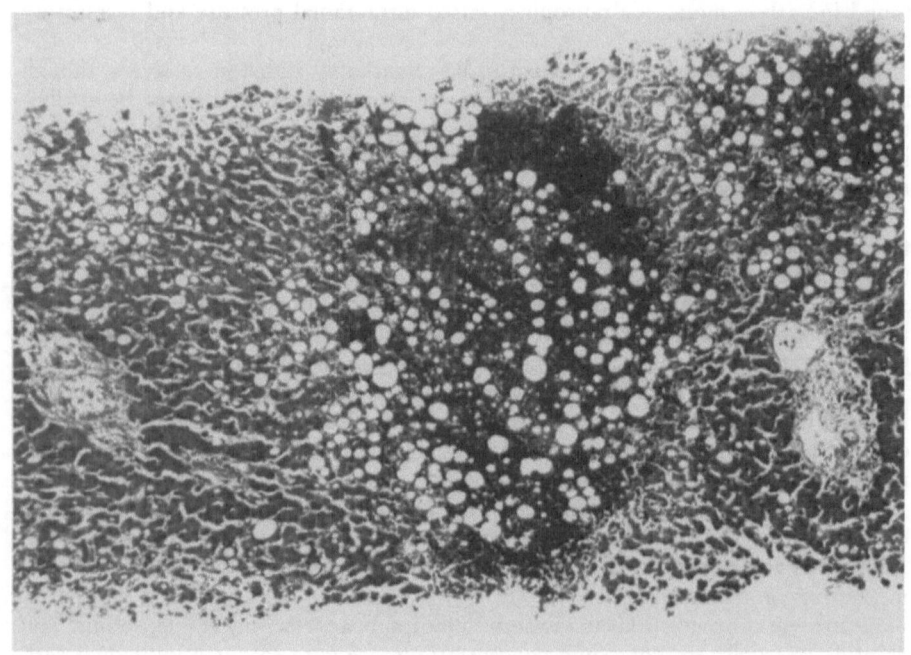

Fig. 13.1. Needle autopsy of liver. Patient died with deepening jaundice and rising serum transaminases leading to a suspicion that death had resulted from hepatitis; hepatitis B status was unavailable. Histology revealed centrilobular necrosis, congestion and steatosis attributable to cardio-respiratory failure ("ischaemic hepatitis"); with this reassurance a full autopsy was then performed. Haematoxylin and eosin, × 107

feasible as a domiciliary procedure or could be carried out in small hospitals or nursing homes without mortuary facilities.

The largest reported series came from the Kingsbrook Jewish Medical Centre, New York, where 394 neeedle autopsies were performed between 1948 and 1968 on cases where they had failed to get relatives' consent for a full routine autopsy (Wellman 1969). The sampling success rate ranged from 92% for liver to only 34% for kidney. West and Chomet (1957) checked the diagnosis from 50 needle autopsies with the findings from complete routine autopsies; discrepancies were found in over half the cases.

The use of needle techniques to obtain fresh tissue very soon after death was well illustrated by Slack et al. (1973) who, with relatives' consent, used lumbar-puncture needles to aspirate cores of frontal lobe and basal ganglia via the orbital roof. Light and immunofluorescence microscopy were performed and cell cultures established. Two of the four cases examined had Jakob-Creutzfeldt disease; in these circumstances there may be reservations about performing full autopsies.

In summary, the needle autopsy, despite its limitations, deserves to be more widely considered as an alternative to the full routine autopsy. It may be the only way of obtaining tissue sufficiently fresh for detailed microscopy or culture. Sampling limitations are severe, however, and it can never be considered as a substitute for the standard autopsy.

The Frozen Section Autopsy

In an attempt to increase the immediate clinical appeal of the autopsy, McCarthy et al. (1981) cut and examined frozen sections from 50 consecutive autopsies so that a final autopsy report could be issued without further delay. The procedure was subject to the usual errors of interpretation and sampling, though the average number of blocks examined by frozen section was 18 compared with a somewhat extravagant 46 for paraffin sections. Nevertheless, many others have confirmed for themselves the value of frozen sections of specific lesions as an adjunct to the standard autopsy, though the combination of autolysis and suboptimal section quality often means that the diagnosis has to be deferred until paraffin sections have been prepared and examined. The prospect of being able to issue a comprehensive autopsy report including histology with minimum delay has some appeal and could well stimulate a revival of interest in post-mortem examinations amongst clinicians because the cases will still be fresh in their minds.

Post-mortem Histology

The extent to which organs are sampled for histology will be determined by the protocols, if any, in each institution, the time available for tissue processing and examination, and the specific needs of each autopsy. Unless the workload is especially onerous, most pathologists might be expected to take blocks for histology from major organs such as brain, heart, lungs, liver, kidneys, etc. In the absence of grossly evident lesions which might dictate where the block is taken from, their precise origin (e.g lung lobe) should be standardised to facilitate retrospective studies.

The quality of post-mortem histology may be seriously hampered by autolysis; the pancreas is often the most seriously affected. The pathologist should minimise the risk of further damage to the tissues by avoiding the temptation to wash the cut surfaces of organs with water; this is liable to lyse the cells, causing further deterioration in the subsequent histology. Another problem is bacterial overgrowth; if there has been a terminal bacteraemia then the organisms will continue to multiply in the warm internal organs of a refrigerated cadaver, damaging the tissues by the production of lytic enzymes or gas, the latter often producing quite large cysts. If it is absolutely essential to get tissue for histology with as little risk of autolysis as possible then a rapid post-mortem examination may be indicated. It is rarely possible for this to be arranged satisfactorily because permission must be obtained and the mortuary may not be available immediately or there may be a delay in transporting the cadaver. With appropriate consent, however, it is possible to sample the larger organs using needles (see pp. 202–203).

It is also necessary to bear in mind that post-mortem tissues are susceptible to mechanical trauma and they should be handled with appropriate care before and during sampling for histology. Blocks for histology should be cut and trimmed with sharp knives and not with relatively blunt instruments or scissors which may tear the tissues.

When interpreting post-mortem histology certain points need to be borne in mind. Allowance must be made for the fact that the typical appearances seen in biopsies from living patients will be frequently modified by treatment administered during life, agonal changes (e.g. congestion and fatty change resulting from terminal cardiac failure) and autolysis. Also it should be noted that mitoses often proceed to completion after somatic death; thus mitotic figures are infrequently seen in post-mortem tissues and are therefore a relatively unimportant indicator of malignancy in tumours discovered at autopsy.

Immunohistology and electron microscopy are possible on post-mortem tissues. Some antigens and ultrastructural features are remarkably well preserved and the results can be unpredictably good. A needle autopsy performed with the minimum of delay may be the best way of obtaining tissues for these investigations in cases where the need has been anticipated.

References

Anonymous (1981) Mini-necropsy. Lancet II: 1026–1027

Berry CL (1980) The examination of embryonic and fetal material in diagnostic histopathology laboratories. J Clin Pathol 33: 317–326

Cabot RC (1912) Diagnostic pitfalls identified during a study of 3000 autopsies. JAMA 59: 2295–2298

Cameron HM, McGoogan E, Watson H (1980) Necropsy: a yardstick for clinical diagnosis. Br Med J 281: 985–988

Castiglioni AL (1947) A history of medicine, 2nd edn. Translated and edited by Krumblhaar EB. Alfred A Knopf, New York, p 567

Castleman B, Chase RA, Rachnod M (1974) Role of the autopsy in the teaching of gross anatomy. N Engl J Med 291: 1413–1414

Chavarria AP, Shipley PG (1924) The Siamese twins of Espanola. Ann Med Hist 6: 297–302

Coe JI (1969) Postmortem chemistries on human vitreous humor. Am J Clin Pathol 51: 741–750

Coe JI (1974) Postmortem chemistry: practical considerations and a review of the literature. J Forensic Sci 19: 13–32

Cope O (1968) The Endicott House conference on medical education: views of medical education and medical care. Cited by Castleman et al. (1974)

Edelstein L (1967) The history of anatomy in antiquity. In: Temkin O, Temkin CL (eds) Ancient medicine: selected papers. Johns Hopkins Press, Baltimore

Enticknap JB (1960) Biochemical changes in cadaver sera. J Forensic Med 7: 135–146

Goldman L (1984) Diagnostic advances versus the value of the autopsy: 1912–1980. Arch Pathol Lab Med 108: 501–505

Goldman L, Saylon R, Robbins S et al. (1983) The value of the autopsy in three medical eras. N Engl J Med 308: 1000–1005

Gruver RH, Freis ED (1957) A study of diagnostic errors. Ann Intern Med 47: 108–120

Hamilton-Paterson JL, Johnson EWM (1940) Postmortem glycolysis. J Pathol Bact 50: 473–482

Hill E (1941) Significance of dextrose and nondextrose reducing substances in postmortem blood. Arch Pathol 32: 452–473

HMSO (1978) Code of practice for the prevention of infection in clinical laboratories and post-mortem rooms. Her Majesty's Stationery Office, London

Jarcho S (1971) Problems of the autopsy in 1670. Bull NY Acad Med 47: 792–796

King LS, Meehan MC (1973) A history of the autopsy: a review. Am J Pathol 73: 514–544

Kottler A (1957) The Jewish attitude on autopsy. NY State J Med 57: 1649–1650

Langley FA (1971) The perinatal postmortem examination. J Clin Pathol 24: 159–169

Leahy MS, Farber ER (1967) Postmortem chemistry of human vitreous humor. J Foresic Sci 12: 214–222

Ludwig J (1972) Current methods of autopsy practice. Saunders, Philadelphia

Lundberg GD (1983) Medical students, truth, and autopsies. JAMA 250: 1199–1200

McCarthy EF, Gebhardt F, Bhagavan BS (1981) The frozen-section autopsy. Arch Pathol Lab Med 105: 494–496

McPhee SJ, Bottles K (1985) Autopsy: moribund art or vital science? Am J Med 78: 107–113

Rother LJ (1966) Rudolf Virchow's views on pathology, pathological anatomy, and cellular pathology. Arch Pathol 82: 197–204

Schmidt S (1983) Consent for autopsies. JAMA 250: 1161–1164

Sigerist HE (1967) Primitive and archaic medicine, Oxford University Press, Oxford

Slack PM, Pryor DS, Dayan AD (1973) Rapid needle sampling of the brain after death for pathology and virology. Lancet I: 521

Terry R (1955) Needle necropsy. J Clin Pathol 8: 38–41

Trexler PC, Gilmour AM (1983) Use of flexible plastic film isolators in performing potentially hazardous necropsies. J Clin Pathol 36: 527–528

Trump BF, Valigorsky JM, Jones RT et al. (1975) The application of electron microscopy and cellular biochemistry to the autopsy: observations on cellular changes in human shock. Human Pathol 6: 499–516

Underwood JCE, Slater DN, Parsons MA (1983) The needle necropsy. Br Med J 286: 1632–1634

Vuori E, Renkonen O-V, Lindbohm R (1983) Validity of post-mortem blood alcohol values. Lancet I: 761–762

Wellman KF (1969) The needle autopsy: a retrospective evaluation of 394 consecutive cases. Am J Clin Pathol 52: 441–444

Wells HG (1923) Relation of clinical to necropsy diagnosis in cancer and value of existing cancer statistics. JAMA 80: 737–740

West M, Chomet B (1957) An evaluation of needle necropsies. Am J Med Sci 234: 554–560

Subject Index